Food and You

A Guide to Modern Agricultural Biotechnology

A publication of the

ACSH American Council on Science and Health

Science. Not Hype.

Written by

Martina Newell-McGloughlin, D.Sc.
Director, UC Systemwide Biotechnology Research and Education Program
University of California, Davis

Bruce Chassy, Ph.D.
Professor Emeritus, Department Food Science and Human Nutrition
University of Illinois at Urbana-Champaign

Gregory Conko, J.D.
Executive Director and Senior Fellow
Competitive Enterprise Institute

Food and You: A Guide to Modern Agricultural Biotechnology.

American Council on Science and Health
1995 Broadway, Suite 202
New York, New York 10023-5860
Tel. (212) 362-7044 • Fax (212) 362-4919
URL: http://www.acsh.org • Email: acsh@acsh.org

Publisher Name: American Council on Science and Health
Title: Food and You: A Guide to Modern Agricultural Biotechnology
Price: $14.95
Authors: Drs. Martina Newell-McGloughlin and Bruce Chassy, and Mr. Gregory Conko
Subject (general): Science and Health
Publication Year: 2013 Binding Type (i.e. perfect (soft) or hardcover): Perfect

ISBN: 978-0-9727094-8-4

Acknowledgements

The American Council on Science and Health (ACSH) appreciates the contributions of the reviewers named below:

Kent Bradford, Ph.D.
Distinguished Professor of Plant Sciences
Director, Seed Biotechnology Center
University of California, Davis

James I. Burke, Ph.D.
Professor and Chair of Crop Science
University College Dublin

Christopher J. Leaver, Ph.D.
Fellow of St John's College
University of Oxford

Alan G. McHughen, D.Phil.
CE Plant Biotechnologist
University of California, Riverside

Table of Contents

Abstract

New and innovative techniques will be required to improve the production and efficiency of the global agriculture sector to ensure an ample supply of healthy food. This challenge is confounded by the inequity between the affluent and developing countries that is likely to continue to widen. It appears that only a handful of technologies are affordable by the least developed countries and are sufficiently scale neutral to be accessible to poorer countries. Biotechnology is one such technology that offers efficient and cost-effective means to produce a diverse array of novel, value-added traits and products.

The first biotechnology products commercialized in agriculture were crops with improved agronomic traits, primarily pest disease resistance and herbicide tolerance whose value was of benefit to the farmers but often opaque to consumers. Currently under development are crops with a more diverse set of new traits that can be grouped into four broad areas, each presenting what, on the surface, may appear as unique challenges and opportunities.

The present and near-future focus is on 1) continuing improvement of agronomic traits such as yield and abiotic stress resistance in addition to the biotic stress tolerance of the present generation of crops; 2) optimizing crop plants for use as biomass feedstocks for biofuels and "bio-synthetics"; 3) the introduction of value-added output traits such as improved nutrition and food functionality; and 4) the application of plants as production factories for therapeutics and industrial products. From a consumer perspective the focus on value added traits, especially improved nutrition, is undoubtedly one of the areas of greatest interest.

Introduction

1

The introduction of *in vitro* methods for the transfer of genes (or other DNA sequences) from one organism to another is a powerful genetic tool that has been applied in medicine, agriculture, and the food and chemical industries. The application of genetic engineering techniques, sometimes called the *new biotechnology*, has led to innovations as varied as the production of new life-saving pharmaceuticals and crops that do not require the use of synthetic insecticides to control insect pests. This paper focuses on the present and future application of genetic engineering techniques for the improvement of agriculture. The introduction of these new molecular techniques represents a paradigm shift in agriculture. As is often the case when new technologies are introduced, concerns have been raised about the safety of novel organisms produced using biotechnology and new risks that could be associated with their introduction into the environment.

The following paragraphs briefly review the development of the food and agricultural system, and the need for major increases in global agricultural productivity. They also discuss the nature of the changes that occur in plant breeding and describe the potential risks of the new technology from a scientific perspective.

Throughout history, paradigm shifts can often be traced to a convergence of events where chance favors the prepared mind or, in the case of the history of technology, prepared collective minds. Research from the divergent disciplines of molecular evolution and archeology support the conclusion that one of the most significant convergences in the history of modern civilization occurred in the marshlands created by the Tigris and Euphrates rivers designated by historians as Mesopotamia and largely occupying the modern region of southern Iraq – the "fertile crescent." The world generally credits the Sumerians, who lived in this region, with the development of civilization. Although nearly contemporary river valley civilizations also developed in the Nile Valley of Egypt and the Indus Valley of Pakistan, the Sumerians seem to have been the first people to live in cities and to create a system of writing (Whitehouse, 1977). Scientists also regard the "fertile crescent," an arc linking Iran, Iraq, Syria, Lebanon, Jordan and Israel/Palestine, as the site of the earlier "Neolithic revolution," when hunter-gatherers first learned to plant crops, then created permanent settlements to cultivate, guard and harvest them. One line of evidence that supports this conclusion is that wild ancestors of the food crops associated with traditional Middle Eastern and European agriculture are native to this region. The general consensus among historians and anthropologists is that by providing a reliable source of food energy that could be stored for long periods of time, carbohydrates were the principle trigger for this birth of civilization. Cereal grasses of this region have long been considered among the first cultivated crops. Adoption of grain crop cultivation has been considered to be a prerequisite for both eastern and western civilization as settled communities required structures in place to manage land and other resources.

In the intervening 10,000 years, many technologies have been developed and used to enhance productivity of that original coterie of cultivated crops and to bring more into the domestic fold. In the latter half of the 20th Century, major improvements in agricultural productivity were largely based on selective breeding programs for plants and animals, intensive use of chemical fertilizers, pesticides and herbicides, advanced equipment developments and widespread irrigation programs. This has been a very successful model for raising productivity but is not without cost and consequent unsustainable damage to the environment.

In addition, from a global perspective, these advances have been the prerogative of more affluent regions. Farmers in developing countries have not had access to many of these technologies and the more capital-intensive methods of improved production (FAO, 2012). During the coming decades, food and agricultural production systems must be significantly enhanced to respond to a number of wide-ranging and far reaching transformations, including a changing and often unpredictable climate, growing world population, increasing international competition, globalization, shifts to increased meat consumption in developing countries and rising consumer demands for improved food quality, safety, health enhancement and convenience (UK, 2012). In order to feed a predicted population of roughly 9 billion by 2050, the world will have to double its annual agricultural production over the next 25 years, despite having already quadrupled it in the last 50 years. The inequities between the affluent and developing countries must be addressed using technologies that are scalable across these economic imbalances. Of even greater concern is the very immediate state of current global food reserves. In 2012 the United Nations issued an unprecedented warning about the state of global food supplies (Eliasson, 2012). They noted that failing harvests in the United States, Ukraine and other countries in 2012 eroded global food reserves to their lowest level since 1974, when the world's population was much lower. World grain reserves are so dangerously low that another year of severe weather in the United States or other food-exporting countries could trigger a major hunger crisis. Clearly, unprecedented needs require innovative solutions to ensure an ample supply of healthy food against competing interests, and this can only be achieved by improving the effectiveness of all components of the global agriculture sector. Innovation is essential for sustaining and enhancing agricultural productivity, and this involves new, science-based products and processes that contribute reliable methods for improving quality, productivity and environmental sustainability.

Agriculturalists have applied approaches such as cross breeding, mutation selection and culling those with undesirable characteristics to modify animals and crop plants over the millennia (Chrispeels and Sadava, 2003). All of these methods depend directly on the selection of desirable novel traits that arise from a variety of kinds of DNA mutations; said another way, novel traits are the results of genetic changes. Thus, from a scientific perspective the term "genetically modified organism" is not an accurate descriptor solely of the products of modern biotechnology, as virtually all domesticated crops and animals have been subjected to varying degrees of genetic modification and selection. Over time, and especially during the last century, plant and animal breeders expanded the tools of genetic manipulation beyond conventional cross breeding to use a variety of other breeding techniques. In the case of plants, these tools include aneuploidy, polyploidy, embryo rescue, protoplast fusion, somaclonal variation, anther culture, colchicine for chromosome doubling and mutation breeding through the use of either radiation or chemicals (Chrispeels and Sadava, 2003).

Crops developed using the methods described above are common throughout the food chain. For example, seedless banana and watermelon varieties were developed using aneuploidy to triple the number of chromosomes. Bread wheat, developed thousands of years ago, is an allopolyploid plant, containing six entire sets of chromosomes from three different species. Broccoflower was developed using embryo rescue, and male sterility in cauliflower was produced by fusing together protoplasts of radish and cauliflower. Many common tomato varieties are the result of wide crosses between domesticated tomato and wild relatives known to contain high levels of glycoalkaloid toxins. Common varieties of Asian pear, grapefruit and durum pasta wheat were developed with irradiation, or "mutation" breeding for fungal resistance in the former and modified starch in the latter (Newell-McGloughlin, 2008).

Innovations such as these have been essential for sustaining and enhancing agricultural productivity in the past and will be even more important in the future. Innovation is necessary to develop new, science-based products and processes that contribute reliable methods for improving quality, productivity and environmental sustainability. Biotechnology has introduced a new dimension to such innovation, offering efficient and cost-effective means to produce a diverse array of novel, value-added products and tools. It has the potential to improve qualitative and quantitative aspects of food, feed and fiber production, reduce the dependency of agriculture on chemicals (a transition from "chemical solutions to biological solutions") and slow the increase in the cost of raw materials, all in an environmentally sustainable manner.

Many of the products we eat or wear can be made using the tools of biotechnology. It is possible to enhance the nutritional content, texture, color, flavor, growing season (time to flowering), yield, disease or pest resistance and other properties of production crops. Transgenic techniques can be applied to farmed animals to improve their growth, fitness and other qualities. Enzymes produced using recombinant DNA methods (in microorganisms/bacteria and yeasts/fungi, etc.) are used to make cheese, keep bread fresh, produce fruit juices and wines and treat fabric for blue jeans and other denim clothing. Other recombinant DNA enzymes are used in laundry and automatic dishwashing detergents.

We can also engineer recombinant microorganisms to improve the quality of our environment. In addition to the opportunities for a variety of new products, including biodegradable products, bioprocessing using engineered microbes offers new ways to treat and use wastes and to use renewable resources for materials and fuel. Instead of depending on non-renewable fossil fuels, we can engineer organisms to convert maize and cereal straw, forest products and municipal waste and other biomass to produce fuel, plastics and other useful commodities. Naturally-occurring microorganisms are being used to treat organic and inorganic contaminants in soil, groundwater and air by a process known as bioremediation. This application of biotechnology has created an environmental biotechnology industry important in water treatment, municipal waste management, hazardous waste treatment, bioremediation and other areas.

Used effectively, biotechnology has enormous potential to improve the quality of our life and our environment. This paper focuses a set of technologies that can be grouped together under the rubric *modern biotechnology* or *new biotechnology*. It explains briefly what these tools are and how these tools are used to improve crops. The present day applications of these new technologies are described with particular emphasis on transgenic crops – often referred to as genetically modified organisms (GMOs) or genetically-engineered (GE) organisms. Current research and future developments are highlighted. Regulation, public acceptance and barriers to adoption round out the discussion.

This paper provides an overview of the tools, techniques and processes that comprise modern molecular biotechnology, with a primary focus on the genetic modification of plants. It is not meant to be a comprehensive review of all the available technologies or the pros and cons of their current applications, but it is intended to offer a guide, or reference work, beyond the mere basics. Section 2 introduces many of the technologies that have been developed for precisely modifying plants at the molecular level, including both those routinely used by plant genetic engineers and a few whose use has only recently been introduced in this field and others whose application will continue to evolve. Section 3 provides a deeper look at the safety of modified plants for the environment and for use in food and livestock feed. Importantly, it puts the safety of genetically engineered plants into the context of what we've learned over thousands of years of plant genetic modification, and it explains why plant biologists are so confident in the potential benefits of the newer technologies.

Section 4 reports on the types of traits that have been introduced into crop plants using recombinant DNA techniques, and it offers a preview of traits that have been or are now being developed. Section 5 summarizes how the current generation of biotech crop plants is being deployed and the agronomic, economic and environmental benefits that growers have reaped from these products, while also anticipating benefits that may be enjoyed from the next generation of biotech varieties. Sections 6, 7 and 8 look at downstream benefits – those already seen as well as those projected – for consumers, such as new sources of energy and fuels, reduced loss of foodstuffs after harvest to spoilage and pests and increased nutritional value.

Section 9 examines some of the hurdles biotech crop developers, seed breeders and farmers have faced when introducing new varieties, while section 10 looks at ways of ensuring peaceful coexistence between growers who choose biotech crop varieties and those who wish to remain "GE-free." On the latter issue, there is considerable discussion as to the most appropriate terminology to use for the general class of the products of modern molecular biological techniques, primarily recombinant DNA technology, with the terms genetically modified organisms, GMOs, biotech crops, etc., used somewhat interchangeably. We will primarily use GE.

This paper does not discuss in detail agricultural and environmental risk assessment and regulation. The impact of GE crops on agriculture, agricultural sustainability and the environment are briefly described in the text.

2 The Technologies

In the simplest and broadest sense, biotechnology is a series of enabling technologies that involve the manipulation of living organisms or their subcellular components to develop useful products and processes. The capacity to manipulate the genetic makeup of living organisms with complexity and precision has become one of the cornerstones of modern biotechnology. It enables developers to enhance the ability of an organism to produce a particular chemical product (e.g., penicillin from a fungus), to prevent it from producing a product (e.g., ethylene in plant cells) or to enable it to produce an entirely new product (e.g., chymosin in microorganisms).

Most of the fundamental technologies that fall within the broad rubric of biotechnology are well known. The most prevalent include recombinant DNA technology; high throughput sequencing; DNA microarrays; RNA interference; genomics, proteomics, metabolomics and bioinformatics; and derivative technologies that use the tools of biotechnology, such as marker-assisted selection and novel haploid generation. Before focusing on these technologies, a brief description of a few of the more fundamental tools of biotechnology research is appropriate.

Recombinant DNA technology

The steps involved in recombinant DNA technology are: (1) to identify the gene that directs the production of the desired substance, (2) to isolate the gene using restriction endonucleases/enzymes, (3) to insert the gene with appropriate regulatory DNA sequences that control expression of the gene into a suitable DNA molecule (vector) for transformation and (4) to transfer the recombined DNA into the appropriate host organism, generally by transforming dozens of single cells in culture. The final step is (5) to select, using a selectable marker gene incorporated in the vector, the transformed cell or cells that have the most desirable characteristics and propagate one or more whole plants from those cells.

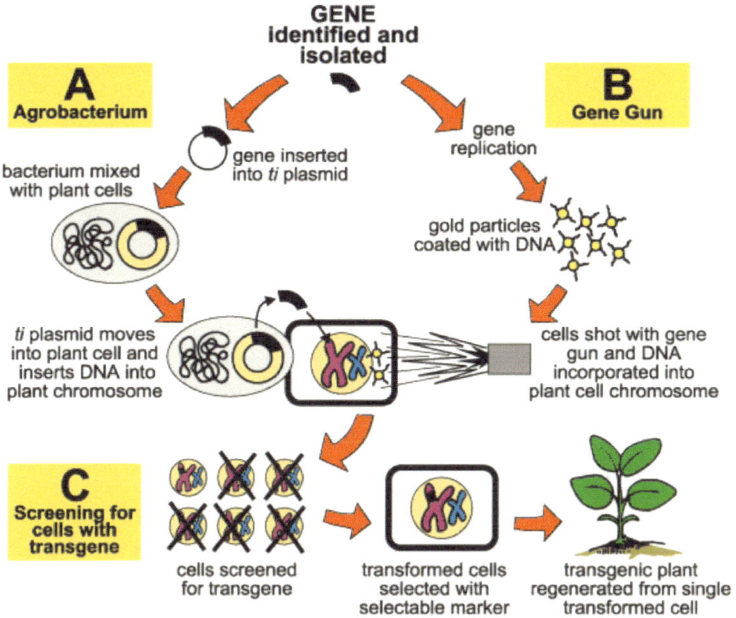

Plant transformation is typically accomplished by using either *Agrobacterium tumefaciens* or a gene gun (see Figure). *Agrobacterium tumefaciens* is a bacterium that occurs in nature and which causes crown gall disease in many species of plants. It contains a small circular piece of non-chromosomal DNA called a Ti plasmid (Ti for tumor inducing). When this bacterium infects susceptible plants, the Ti plasmid enters cells of the host plant, and specific regions of the Ti plasmid insert themselves into the host cell's genome. This insertion occurs in a region of the DNA strand with a specific sequence. The host cell then expresses the gene from the bacteria, which induces massive cell growth and the plant tumors after which the bacterium is named.

Biotechnology utilizes this natural transformation process by removing the bacterial genes from the region transferred to the host genome and substituting the gene(s) of interest (see part A of Figure). Agrobacterium use for transformation was initially limited to use in certain dicotyledonous, or broad-leaf, plants because the wild-type microbe tends only to infect those species. The system has since been improved in the laboratory to allow agrobacterium-mediated transformation of most major crops, including monocotyledonous, or "grassy," plants. But agrobacterium is still mainly used to transform dicots.

The other transformation process involves coating tiny gold, tungsten or other heavy metal particles with genes of interest. The coated particles are shot into single cells of the plant of interest using compressed air or another propellant. This is commonly referred to as particle bombardment/acceleration, biolistics or the gene gun approach. In a process not fully understood, the transgene(s) are incorporated into a DNA strand of the host genome (see part B of Figure). This process is inefficient but does not have the host species limitation of agrobacterium.

After transformation with either the agrobacterium or gene gun processes, untransformed cells must be eliminated. This is facilitated with the use of selectable marker genes. In the case of an herbicide tolerance gene, the herbicide tolerance trait itself serves as the selectable marker, since the herbicide will kill non-transformed cells (see part C of Figure). When another trait of interest is being transformed in the crop, a selectable marker, like antibiotic resistance or herbicide tolerance, is used in addition to the primary gene of interest. The cells in culture are treated with the herbicide or an antibiotic, and only those cells that were transformed with the selectable marker genes will survive. Whole plants are then regenerated from the surviving cells by growth on an appropriate medium containing plant hormones.

Following transformation and plant regeneration, the transgenic plants are first extensively tested in the laboratory under containment conditions before extensive testing in field in a range of geographical locations to ensure that the transgene functions properly and confers the desired trait. Not all transgenic plants will express the trait or gene product properly or stably, and these are eliminated before field testing. Once a transgenic plant with robust and stable trait expression has been identified, the trait can then be bred using conventional plant breeding methods into elite crop cultivars best suited to the environmental conditions where the crop is grown.

Microarrays (biochips)

DNA microarrays, also commonly known as biochips or DNA chips, were developed in the late 1980s and early 1990s primarily by the company Affymetrix. Biochips are a "massively parallel" genomic technology. They facilitate high throughput analysis of thousands of genes simultaneously and are thus a potentially very powerful tools for gaining insight into the complexities of higher organisms including analysis of gene expression, detecting genetic variation, making new gene discoveries, "fingerprinting strains" and developing new diagnostic tools. These technologies permit scientists to conduct large-scale surveys of gene expression in organisms, thus adding to our knowledge of how they grow and develop over time or respond to various environmental conditions. These techniques are especially useful in gaining an integrated view of how multiple genes are expressed in a coordinated manner. This technology is now largely superseded by next generation sequencing.

RNA silencing

Increased production of desirable characteristics present in crops is a common goal for breeders – for example, grains with increased protein content and nutritional quality, fruits and vegetables with enhanced nutritional value and flowers with deeper colors. It was in pursuit of the latter goal that a most bemusing and ultimately valuable phenomenon was first observed. While attempting to create "black" petunias as a model for one of the ultimate floriculture aspirations, the "black" rose, Jorgensen et al. (1996) of UC Davis attempted to over-express the chalcone (a pigment precursor) synthesis gene by introducing a modified copy under a strong promoter. The surprising result was white flowers, and many strange variegated variations between purple and white. This was the first demonstration of what has come to be known as post-transcriptional gene silencing (PTGS). While initially it was considered a strange phenomenon limited to some plant species, it is now recognized to be a significant regulatory mechanism in all higher organisms. RNA interference (RNAi) in animals and basal eukaryotes, "quelling" in fungi and posttranscriptional gene silencing (PTGS) in plants are examples of a broad family of phenomena collectively called RNA silencing (Hannon 2002; Plasterk 2002). In addition to its occurrence in these organisms, it has roles in viral defense and transposon silencing mechanisms, among other things. Perhaps most exciting, however, is the emerging use of PTGS and, in particular, RNA interference (RNAi) – PTGS initiated by the introduction of double-stranded RNA (dsRNA) – as a tool to "knock out" expression of specific genes in a variety of organisms.

Instead of producing large quantities of new proteins, high-expressing transgenes (genes from another source) introduced into the plant can actually inhibit the expression of the plant's own genes by triggering sequence-specific destruction of similar transcripts (Dykxhoorn et al., 2003). Thus, the transgene ends up silencing both its own expression and that of similar endogenous genes when the concentration of transgene transcript (mRNA) becomes too high in the cytoplasm. This unintended "RNA silencing" can nonetheless be harnessed by scientists, for example to eliminate unwanted gene expression, and is used by the plant itself to inhibit protein synthesis by infecting RNA viruses. A significant example of this (which was determined after the fact) is coat-protein-gene protection introduced into papaya to protect against papaya ring spot virus, one of the few transgenic gene-silencing systems in commercial production (Fitch et al., 1992). Recent studies have demonstrated that RNAi-mediated mechanisms and PTGS have been unknowingly exploited for many years by plant breeders in many new crop phenotypes (Parrott et al., 2010). Parrott et al. concluded that crops modified using small RNA technology that produce no novel proteins or metabolites pose little if any novel hazards to food safety or the environment and that a subset of the currently applied regulatory paradigm can be used for their safety assessment.

Riboswitches

Each cell must regulate the expression of hundreds of different genes in response to changing environmental or cellular conditions. The majority of these sophisticated genetic control factors are proteins, which monitor metabolites and other chemical cues by selectively binding to targets. RNA also can form precision genetic switches and these elements can control fundamental biochemical processes.

Riboswitches are a type of natural genetic control element that uses an untranslated sequence in an mRNA to form a binding pocket for a metabolite that regulates expression of that gene. Potentially engineered riboswitches might function as designer genetic control elements.

Protein engineering

Another area of genetic engineering, in the broader sense, is protein engineering (Brannigan and Wilkinson, 2002). New enzyme structures may be designed and produced in order to improve on existing enzymes or create new activities. The principal approaches are: 1) site-directed mutagenesis (oligonucleotide-directed mutagenesis), 2) random mutation and selection and 3) directed evolution, which is a refinement of the latter. However, from a practical point of view, much of the research effort in protein engineering has gone into studies concerning the structure and activity of enzymes chosen for their theoretical importance or ease of preparation rather than industrial relevance. With a greater focus on "green" production systems, this emphasis is now shifting.

Genomics, proteomics, metabolomics and bioinformatics

Functional genomics can be defined as establishing a link between gene expression and cellular function (Prevsner, 2009). Thousands of genes in a single tissue type vary in levels of expression at different developmental stages, in health and disease, at different chronological times and in response to environmental variation. Functional genomics provides insight into all the genes involved and the roles they play. Although applications in medicine have been the main thrust of this development, this technology is now having a major impact in agricultural biotechnology, and most specifically plant biotechnology research.

The proteome is the complete set of proteins expressed and modified after their expression from the genome. Proteomics refers to the study of the structure, function, location and interaction of proteins within and between cells. Proteomics is particularly being driven by global analyses of gene expression and inferences derived from DNA sequence data. However, the study of proteins is not a simple linear extrapolation from knowledge of the DNA sequence. It is a highly complex multidimensional field of endeavor. Each cell produces thousands of proteins, each with a specific function. Proteins differ greatly from one another, even within the same individual, but DNA molecules are remarkably similar. In addition, unlike the unvarying genome, an organism's proteome is so dynamic that an almost infinite variety of protein combinations exist. The proteome varies from tissue to tissue, cell to cell, and with age and time. The cellular proteome changes in response to other cells in the body and to external environmental stimuli. Methods for analysis of protein profiles and cataloging protein-protein interactions on a genome-wide scale are technically more difficult but are improving rapidly, especially for microbes. Functional genomics will impact most areas of biology, from fundamental biochemistry to improvement of quality, and agronomic traits in crops, improved protection against pathogenic microbes, and improved exploitation of beneficial microbes.

Sequencing of organismal genomes has created a vast quantity of data, which is not easily examined or understood. In many ways plant sequences are more complex than even the human genomes given the sheer amount of DNA. Similar problems exist for a wide variety of topics in functional genomics, proteomics, metabolomics and so on, primarily due to the scale and parallel nature of these approaches.

The construction of relational databases, as well as the development of efficient methods for searching and viewing these data, constitutes a discipline called "bioinformatics." In a broader view, bioinformatics contains computational or algorithmic approaches to the production of information from large amounts of biological data, and this might include prediction of protein structure, dynamic modeling of complex physiological systems or the statistical treatment of quantitative traits in populations to determine the genetic basis for these traits. Unquestionably, bioinformatics is now an

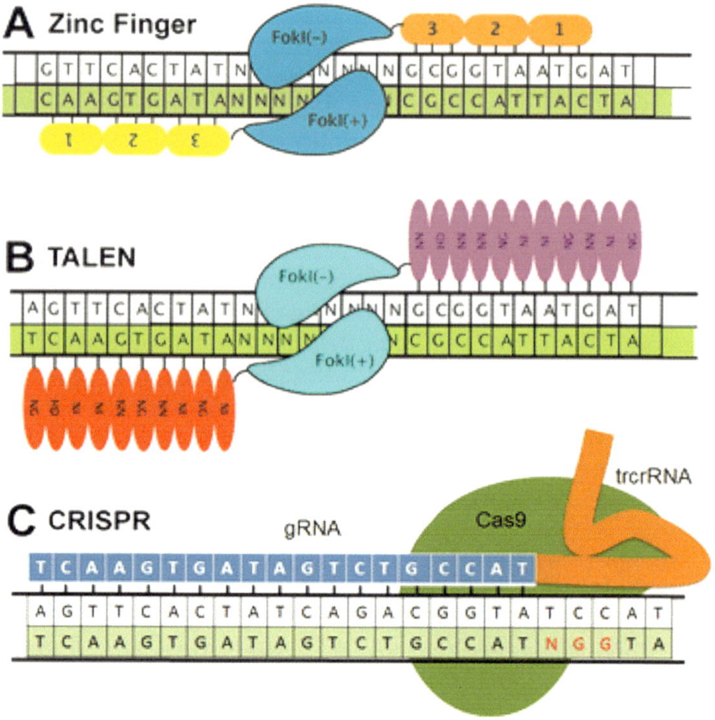

essential component of all research activities utilizing structural and functional genomics approaches for analysis at the sequence level, in structure modeling, and in modeling, linking and simulating complex higher-level structures such as metabolic and neurological pathways.

Evolving techniques

As with any rapidly evolving field tools and techniques are being improved at a rapid pace. A brief overview will be provided here. This not intended to be either comprehensive or complete review but serves as an example of the rapid development of new molecular tools.

Genome editing

Figure 2. Genome editing tools (source: Addgene https://www.addgene.org/CRISPR/guide/)

One of the subtler modification systems, which will provide a challenge for regulatory oversight, is what may be referred to as next generation directed mutagenesis. Genome editing, specifically genome editing with engineered nucleases (GEEN), is a form of fined-tuned gene modification using novel nucleases which results in minor insertions, replacements or deletions in a highly targeted manner. There are a number of gross categories of engineered nucleases, mostly based, like the original restriction endonucleases, on immune systems from their source organisms: Zinc Finger Nucleases (ZFNs), Transcription Activator-Like Effector Nucleases (TALENs), Clustered Regularly Interspaced Short Palindromic Repeats (CRISPRs) and homing endonucleases including engineered hybrid meganucleases. These nucleases create targeted specific double-stranded breaks at desired locations in the genome and harness the cell's endogenous mechanisms to repair the induced break by natural processes of homologous recombination (HR) and non-homologous end-joining (NHEJ) (Esvelt and Wang, HH., 2013).

Genome editing with these nucleases is different from silencing the gene of interest by short interfering RNA (siRNA) in that the engineered nuclease is able to modify DNA-binding specificity and therefore can, in principle, cut any targeted position in the genome and introduce modification of the endogenous sequences for genes that are impossible to specifically target by conventional RNAi. These and evolving editing tools have tremendous potential to introduce very targeted modifications, which will call into question the notion of "genetic engineering" and will present a problem for regulatory authorities, if they, without justification, decide they need to capture them for review. It will be, for all intents and purposes, impossible to detect, so enforcing oversight will be prove to be a challenge.

Other systems, such as synthetic biology (the design and construction of new biological parts, devices and systems) and genome-scale engineering are being enabled through advances in high-throughput DNA sequencing, and large-scale biomolecular modeling of metabolic and signaling networks will contribute to food and agriculture production systems in the future.

Metabolic engineering

One field of great interest is the modification of complex traits – traits not necessarily associated with a single inserted gene – especially those that may have an epigenetic component.

Plants produce between 200 and 250,000 secondary metabolites. Analysis of these metabolites (most specifically metabolomic analysis) is a valuable tool in better understanding what has occurred during crop domestication (lost and silenced traits) and in designing new paradigms for more targeted crop improvement that is better tailored to current needs. In addition, with modern techniques, it confers the potential to seek out, analyze and introgress traits of value that were limited in previous breeding strategies. Research to improve the nutritional quality of plants has historically been limited by a lack of basic knowledge of plant metabolism and the challenge of resolving complex interactions of thousands of metabolic pathways. Metabolic engineering is generally defined as the redirection of one or more reactions (enzymatic and otherwise) to improve the production of existing compounds, produce new compounds or mediate the degradation of undesirable compounds. It involves the redirection of cellular activities by the modification of the enzymatic, transport and/or regulatory functions of the cell. Significant progress has been made in recent years in the molecular dissection of many plant pathways and in the use of cloned genes to engineer plant metabolism. So, in sum, a complementarity of techniques both traditional and novel is needed to metabolically engineer plants to produce desired traits.

Although progress in dissecting metabolic pathways and our ability to manipulate gene expression in genetically-engineered (GE) plants has progressed apace, attempts to use these tools to engineer plant metabolism have not quite kept pace. Since the success of this approach hinges on the ability to change host metabolism, its continued development will depend critically on a far more sophisticated knowledge of plant metabolism, especially the nuances of interconnected cellular networks, than currently exists. This complex interconnectivity is regularly demonstrated. Relatively minor genomic changes (point mutations, single-gene insertions) are regularly observed following metabolomic analysis to lead to significant changes in biochemical composition. However, what on the surface would appear to be other, more significant genetic changes unexpectedly yield little phenotypical effect.

A number of new approaches are being developed to counter some of the complex problems in metabolic engineering of pathways. Such approaches include use of RNA interference to modulate endogenous gene expression or the manipulation of transcription factors (Tfs) that control networks of metabolism (Bruce et al., 2000; Gonzali, Mazzucato, & Perata, 2009; Kinney AJ, 1998). For example, expression in tomatoes of two selected transcription factors (TFs) involved in anthocyanin production in snapdragon (*Antirrhinum majus L.*) led to accumulation of high levels of these flavonoids throughout the fruit tissues, which, as a consequence, were purple. Such expression experiments hold promise as an effective tool for the determination of transcriptional regulatory networks for important biochemical pathways (Gonzali S et al., 2009). Gene expression can be modulated by numerous transcriptional and posttranscriptional processes. Correctly choreographing the many variables is the factor that makes metabolic engineering in plants so challenging.

In addition, there are several new technologies that can overcome the limitation of single gene transfers and facilitate the concomitant transfer of multiple components of metabolic pathways. One example is multiple-transgene direct DNA transfer, which simultaneously introduces all the components required for the expression of complex recombinant macromolecules into the plant genome as demonstrated by a number including Nicholson et al. (2005), who successfully delivered four transgenes that represent the components of a secretory antibody into rice (Carlson et al., 2007), constructing a minichromosome vector that remains autonomous from the plant's chromosomes and stably replicates when introduced into maize cells. This work makes it possible to design minichromosomes that carry cassettes of genes, enhancing the ability to engineer plant processes such as the production of complex biochemicals.

It was demonstrated (Christou P, 2005) that gene transfer using minimal cassettes is an efficient and rapid method for the production of transgenic plants stably expressing several different transgenes. Since no vector backbones are present, this technique allows the construction of transformants that contain only the DNA sequences required to produce the desired new trait. They used combinatorial direct DNA transformation to introduce multi-complex metabolic pathways coding for beta carotene, vitamin C and folate. They achieved this by transferring five constructs controlled by different endosperm-specific promoters into white maize. Different enzyme combinations show distinct metabolic phenotypes resulting in 169-fold beta carotene increase, six times the amount of vitamin C and a doubling of folate production, effectively creating a multivitamin maize cultivar (Naqvi et al., 2009). This system has an added advantage from a commercial perspective in that these methods circumvent problems with traditional approaches, which not only limit the amount of sequences transferred but may disrupt native genes or lead to poor expression of the transgene, thus reducing both the numbers of transgenic plants that must be screened and the subsequent breeding and introgression steps required to select a suitable commercial candidate.

"Omics"-based strategies for gene and metabolite discovery, coupled with high-throughput transformation processes and automated analytical and functionality assays, have accelerated the identification of product candidates. Identifying rate-limiting steps in synthesis could provide targets for modifying pathways for novel or customized traits. Targeted expression will be used to channel metabolic flow into new pathways, while gene-silencing tools will reduce or eliminate undesirable compounds or traits or switch off genes to increase desirable products (Davies, 2007; Herman, Helm, Jung, & Kinney, 2003; Liu, Singh, & Green, 2002). In addition, molecular marker-based breeding strategies have already been used to accelerate the process of introgressing trait genes into high-yielding germplasm for commercialization. Table 1 summarizes the work done to date on specific applications in the categories listed above.

More specific example of technology that applies for complex modifications is described in the relevant applications below.

Table 1 – Examples of Crops in Research with Nutritionally Improved Traits[1]

Trait	Crop (trait detail)	Reference
Protein and amino acids		
Protein quality and level	Bahiagrass (protein↑)	Luciani et al. 2005
	Canola (amino acid composition)	Roesler et al. 1997
	Maize (amino acid composition; protein↑)	Cromwell 1967, 1969;Yang et al. 2002; O'Quinn et al. 2000; Young et al. 2004
	Potato (amino acid composition; protein↑)	Chakraborty et al. 2000; Li et al. 2001; Yu and Ao 1997; Atanassov et al. 2004
	Rice (protein↑; amino acid)	Katsube et al. 1999
	Soybean (amino acid balance)	Rapp 2002; Dinkins et al. 2001
	Sweet Potato (protein↑)	Prakash et al. 2000
	Wheat (protein↑)	Uauy et al. 2006
Essential amino acids	Canola (lysine↑)	Falco et al. 1995
	Lupin (methionine↑)	White et al. 2001
	Maize (lysine↑; methionine↑)	Agbios 2006; Lai and Messing 2002
	Potato (methionine↑)	Zeh et al. 2001
	Sorghum (lysine↑)	Zhao et al. 2003
	Soybean (lysine↑; tryptophan↑)	Falco et al. 1995; Galili et al. 2002
Oils and Fatty Acids		
	Canola (lauric acid↑; γ-linolenic acid↑; + ω-3 fatty acids; 8:0 and 10:0 fatty acids↑; lauric + myristic acid↑; oleic acid↑)	Del Vecchio 1996; Froman and Ursin 2002; James et al. 2003; Ursin, 2003, Dehesh et al. 1996; Agbios 2006; Roesler et al. 1997
	Cotton (oleic acid↑; oleic acid + stearic acid↑)	Chapman et al. 2001; Liu et al. 2002
	Linseed (+ ω-3 and- 6 fatty acids)	Abbadi et al. 2004

[1] Excludes protein/starch functionality, shelf life, taste/aesthetics, fiber quality and allergen reduction traits. Modified from Newell-McGloughlin,2008 (25)

Trait	Crop (trait detail)	Reference
	Maize (oil↑)	Young et al. 2004
	Oil Palm (oleic acid↑ or stearic acid↑; oleic acid↑ + palmitic acid↓)	Parveez 2003; Jalani et al. 1997
	Rice (α-linolenic acid↑)	Anai et al. 2003
	Soybean (oleic acid↑; γ-linolenic acid↑Stearidonic Acid↑)	Kinney and Knowlton 1998; Reddy and Thomas 1996, SDA, 2011
	Safflower (γ Linoleic Acid GLA↑)	Arcadia, 2008
Carbohydrates		
Fructans	Chicory, (fructan↑; fructan modification)	Smeekens 1997; Sprenger et al. 1997 Sévenier et al (1998)
	Maize (fructan↑)	Caimi et al. 1996
	Potato (fructan↑)	Hellwege et al. 1997
	Sugar beet (fructan↑)	Smeekens 1997
Frustose, Raffinose, Stachyose	Soybean	Hartwig et al 1997
Inulin	Potato (inulin↑)	Hellwege et al. 2000
Starch	Rice (amylase ↑)	Chiang et al. 2005, Schwall, 2000
Micronutrients and functional Metabolites		
Vitamins and Carotenoids	Canola (vitamin E↑)	Shintani and DellaPenna 1998
	Maize (vitamin E↑ ; vitamin C↑; beta-carotene↑; folate↑)	Rocheford 2002; Cahoon et al. 2003; Chen et al. 2003; Naqvi et al. 2009
	Mustard (+β-carotene)	Shewmaker et al. 1999
	Potato (β-carotene and lutein↑)	Ducreux et al. 2005
	Rice (+ β-carotene)	Ye et al. 2000
	Strawberry (vitamin C↑)	Agius et al. 2003
	Tomato (folate↑; phytoene and β-carotene↑; lycopene↑; provitamin A↑)	Della Penna, 2007, Díaz de la Garza et al. 2004; Enfissi et al. 2005; Mehta et al. 2002; Fraser et al. 2001; Rosati 2000

Trait	Crop (trait detail)	Reference
Functional 2^{ndry}metabolites	Apple (+stilbenes)	Szanowski et al. 2003
	Alfalfa (+resveratrol)	Hipskind and Paiva 2000
	Kiwi (+resveratrol)	Kobayashi et al. 2000
	Maize (flavonoids↑)	Yu et al. 2000
	Potato (anthocyanin and alkaloid glycoside↓; solanin↓)	Lukaszewicz et al. 2004
	Rice (flavonoids↑; +resveratrol)	Shin et al. 2006; Stark-Lorenzen 1997
	Soybean (flavonoids↑)	Yu et al. 2003
	Tomato (+resveratrol; chlorogenic acid↑; flavonoids↑; stilbene↑anthocynanins↑)	Giovinazzo et al. 2005; Niggeweg et al. 2004; Muir et al. 2001; Rosati, 2000, Gonzali et al, 2009
	Wheat (caffeic and ferulic acids↑; +resveratrol)	UPI 2002
Mineral availabilities	Alfalfa (phytase↑)	Austin-Phillips et al. 1999
	Lettuce (iron↑)	Goto et al. 2000
	Rice (iron↑)	Lucca et al. 2002
	Maize(phytase↑, ferritin↑)	Drakakaki 2005, Han, 2009
	Soybean (phytase↑)	Denbow et al. 1998
	Wheat (phytase↑)	Brinch-Pedersen et al. 2000

Food safety and risk assessment

3

The nature of genetic changes in crops

Historically, agriculturists, and more recently plant breeders, selected improved crops based on changes that arose as a result of genetic modification (naturally occurring mutations) of DNA without any knowledge of the nature of the molecular modifications that had occurred in the DNA or resulting changes in the content of proteins and metabolites contained in newly selected varieties. The introduction of high throughput DNA sequencing methods coupled with bioinformatic analysis, as well as improved methods for evaluating/analyzing the proteome and metabolome of crop plants, has provided insight into the molecular changes that occur as a result of plant breeding. The kinds of DNA modifications that are associated with classical plant breeding or through transgene insertion have been assessed and compared (see for example Parrott et al., 2010; Weber et al., 2012).

There now exists a significant body of evidence that demonstrates that all forms of plant breeding introduce a variety of changes in DNA, ranging from point mutations and single base pair deletions and insertions, loss or acquisition of genes, to changes in numbers of whole chromosomes. When compared to classical plant breeding methods, transgene insertion has been observed to produce less unintended DNA modification. Studies have also shown that transgenic crop varieties more closely resemble their parental lines than do other varieties of the same crop with respect to their proteomic and metabolomics profiles (Ricroch et al., 2012). It should not be overlooked, however, that in most plant breeding programs successive rounds of planting and selection are used to cull out events with obvious and/or undesirable phenotypes and select for plants with unchanged/superior agronomic and phenotypic traits. Regardless of the method of breeding applied to select for genetic changes, the candidate plants that are advanced for potential distribution and planting should closely resemble their parental lines – with the sole exception of the intended modification. It also appears that environmental and cultural conditions have more impact on plant composition than do breeding and selection programs (Ricroch et al., 2012).

In light of these substantial and unpredictable genetic modifications, which occur in crop plants that are common in the human diet, the comparatively simple and more precise modifications performed with recombinant DNA techniques – what we know as modern biotechnology – appear to be unique only in the breeder's improved ability to control the resulting phenotype of the modified cultivars. As long ago as 1987, an analysis published by the National Academy of Sciences (USA) examined the available research and concluded that plants and other organisms produced using genetic engineering techniques pose no new or different risks to human health or the environment than those produced using other breeding methods (NAS, 1987). Since that time, the National Academies, the EU and the governments of a number of countries have on several occasions reviewed the scientific literature on the safety of biotech crops, and each time they have reached the same conclusion.

Transgenic crops produced using the new biotechnology are also regulated by governments and may not be released to farmers or consumers until they have successfully passed a rigorous pre-market safety assessment (Kok and Kuiper, 2003; König et al., 2004; Codex, 2003). On a case-by-case basis, the safety assessor seeks to determine if the new trait introduced into a crop is cause for safety concerns. In principle, the focus of regulators is on the safety of the new trait and not on the fact that genetic engineering has been used to introduce the new trait. Yet, paradoxically, crops developed using less precise and more disruptive methods of breeding may be released without any pre-market regulatory review.

The safety assessor's role is to ensure that the new crop variety is as safe as, or safer than, other varieties of the same crop. Safety review does not, and cannot, prove that a crop is absolutely without risk because, as we discuss below, no food is completely without risk or risk of being used in an unsafe way. As noted above, all breeding produces changes in DNA that could produce unintended hazards (Cellini et al., 2004). There is unfortunately much confusion on this specific issue, with critics of transgenic crops claiming that the use of modern biotechnology is unsafe *per se*. A major source of misunderstanding about the safety of biotech crops stems from the belief that the crops produced around the world today are totally innocuous and that any genetic modification could adversely affect safety.

Are crop plants produced by "conventional" plant breeding absolutely safe to consume? Many crops produce potentially toxic phytochemicals to protect themselves from pests. Potatoes and tomatoes, for example, naturally contain low levels of potentially lethal glycoalkaloid toxins, solanine and chaconine. Rapeseed, the cultivated plant from which canola was derived, contains both toxic erucic acid and anti-nutrient glucosinolates. And kidney beans contain phytohaemagglutinin in levels sufficient to be toxic if undercooked. Cassava (*Manihot esculenta*), a major staple crop in many developing countries, contains cyanogenic glycosides in sufficient quantity to cause death if not properly processed to remove cyanide; chronic low-level consumption can cause goiter. A few of the most commonly eaten plants give rise to food allergies (e.g., peanuts and other groundnuts, tree nuts, soybean, wheat, kiwi fruit and sesame). And an estimated 99 percent or more (by weight) of the pesticides that humans consume in food are chemicals that plants produce naturally to defend themselves from predators. Few of these have been fully tested for human safety. But roughly half of those that have been tested were found to be rodent carcinogens (Ames, Profet, & Gold, 1990).

The natural toxins and other potentially hazardous compounds present in dozens of common food plants can be accidentally raised to harmful levels with basic hybridization plant breeding. And other techniques, such as embryo rescue and mutation breeding could cause unpredictable genetic changes that raise similar risks. These unanticipated effects are, however, less likely to occur with biotechnology breeding. Indeed, biotechnology approaches can be employed to downregulate or even eliminate the genes involved in the metabolic pathways for the production and/or activation of such plant toxins and also allergens such as globulins in peanuts (Dodo, 2008). Ironically, many of our daily staples would be banned if subjected to the rigorous safety and testing standards applied to crop plants modified using recombinant DNA technology.

The older, non-biotech breeding techniques allow for no control at the genome level. Rather, multiple genes are transferred together or mutated and unwanted traits are eliminated through subsequent selection and backcrossing. Plants created by these conventional phenotypic selection techniques undergo no formal food or environmental safety evaluation prior to introduction into the environment and marketplace, other than normal agricultural variety testing. This is not to suggest that classical breeding methods are inherently unsafe. Despite the extensive genetic modification of crop plants by these diverse methods, cases of novel or completely unexpected adverse consequences for commercialized varieties of these crops are extremely rare (e.g., high glycolalkaloid Lenape potato).

Nor should we be concerned that biotechnology methods permit breeders to move genes between unrelated taxonomic kingdoms. The recent massive accumulation of DNA sequencing data shows extensive genetic similarity among genomes of diverse organisms that are only remotely related. For example, parts of the nucleic acid sequence of a common bacterium present in our guts, *Escherichia coli*, have been found in the DNA of organisms such as oilseed rape, amphibians, birds, grasses and mammals – including humans. Such findings put in doubt the value of assigning genes to a particular species and the validity of using terms such as “species-specific” DNA. Due to the common genetic ancestry of all living organisms, there already is broad sharing of identical or very similar genes across taxonomic kingdoms, so there is nothing inherently unnatural or unsafe about moving a gene from bacteria or viruses into a crop plant. Modern biotechnology would not in fact be possible if genes did not function in the same way in all organisms. In particular, the genes that encode proteins that are involved in functions that are common to all cells (for example: DNA replication, RNA transcription and protein synthesis) are similar in all organisms. The safety of any genetic modification depends solely on the function of the specific gene or genes that are moved, how they are expressed in the new host organism and the impact that new phenotype has on the modified organism's environment and use. Genes are not organism specific – harmful genes may be moved with simple hybridization between closely related plant species, and helpful or benign genes may be moved between kingdoms to no ill effect.

Based on the foregoing discussion it should be clear that all new varieties of crops are the result of genetic modification regardless of the technology used for their development. To date, new crop varieties have been almost without exception safe to plant and safe to consume. The small number of documented cases in which a new variety was found to be unsafe for consumers were all the products of classical breeding methods (NRC 2004). Nevertheless, new varieties have proven so comparatively safe that non-biotech ones are released to farmers with essentially no oversight by regulators and, with very few exceptions, no requirements for safety testing. Crops produced using modern biotechnology are, however, all subject to special regulation with associated significant costs implications.

There has been great misunderstanding about why pre-market safety assessment should be required of these crops. It is often asserted that genetic modification may produce unforeseen and unintended changes in crops, but as we have seen all breeding produces unintended changes (see above and Cellini et al., 2004; Parrott, 2010), so that cannot be the scientific basis for regulating these crops. Nevertheless, aspects of the regulatory framework in every country that permits the commercial use of biotech crops, or food and animal feeds derived from them, are premised on the belief that unique risks arise from the transformation process itself. Each time a gene is introduced into a plant, the resulting organism (or "transformation event") is treated as a unique product for the purposes of regulation. Even if copies of a single gene encoding the same protein are inserted into different plants of the same species, each resulting transformation event must be tested and approved separately.

The rationale for such event-specific regulation is that the specific site of transgene insertion into the host genome cannot be targeted and will therefore be unique for each transformant. Because such a random insertion could interfere with the normal functioning of endogenous genes, with potentially harmful unintended effects, regulators presume that each event may be uniquely risky. However, there is no evidence that the uncertainties associated with transgene insertion are any greater than those that occur with other forms of genetic modification, such as the random genetic changes that result from mutation breeding, the pleitropic effects on gene structure and expression common in wide crosses and ploidy modification, or even those that accompany the movement of transposable elements in normal sexual reproduction (Weber et al., 2012; Steiner et al., 2013). Yet, while all breeding methods pose a theoretical risk of unintended genetic changes that could result in an increase in toxins and other harmful substances, or a reduction in dietary nutrients or other beneficial constituents, such effects are routinely identified by basic phenotype analysis. In fact, as Bradford et al. observe, "conventional breeding programs generally evaluate populations with much wider ranges of phenotypic variation than is observed in transgenic programs" (2005, p. 441), and they successfully eliminate potentially harmful plants from development programs. With the advent of inexpensive methods of DNA sequencing, the site of such gene insertion events is now always well characterized.

It is also commonly believed that transgenic crops should be regulated because they express novel traits that not only are not normally associated with that crop, but which in many cases have not been part of the human or animal diet. When a genuinely novel substance (e.g., a new protein or other phytochemical) is introduced into a plant, this does merit special testing to assure the crop is safe for consumers and the environment. But most of the traits introduced into biotech crops currently on the market can also be introduced with various classical breeding methods.

For example, herbicide tolerance, the most widely adopted transgenic trait, is routinely introduced into crop species via selection or mutation breeding. And while the transgene responsible for tolerance to the herbicide glyphosate, a common trait in biotech plants, was isolated from the common soil microbe *Agrobacterium tumefaciens*, the wild type gene and the EPSP synthase protein for which it codes can be detected in many non-transgenic food crops due to the presence of soil residues in harvested crops. Furthermore, all plants in the human diet naturally contain a gene that encodes an EPSP synthase protein that is required for normal plant growth and development. All transgenic insect resistant plants commercialized to date contain a gene from one of several subspecies of naturally occurring soil bacterium *Bacillus thurengiensis* (Bt). But whole Bt spores have long been cultured for use as a "natural" insecticide for food and ornamental crops, so consumers have a long history of exposure to Bt genes and the specific proteins responsible for transgenic insect resistance. In fact, thuricides such as DIPEL made from Bt spores are the only insecticides approved for use by the organic food industry. Moreover, various classical breeding methods, such as interspecific and intergeneric "wide cross" hybridization, frequently introduce new genes and gene products into the human diet. Thus, not all biotech plant varieties contain genes or proteins new to the food supply, nor is the introduction of novel substances unique to transgenic breeding methods.

It is worth repeating that the only scientific justification for pre-market safety assessment for any new plant variety is to establish the safety of any newly introduced substances. It is an unfortunate reality that pre-market safety assessment has become an endless search for unintended effects which have become like the new clothes in fable of the "Emperor's New Clothes." The risk assessment described in the following paragraph has been applied to accomplish the assessment.

Risk assessment

The consensus of scientific opinion and evidence is that the application of GE technology introduces no unique food/feed safety or environmental impact concerns and that there is no evidence of harm from those products that have been through a regulatory approval process. This conclusion has been reached by numerous national and international organizations (e.g., Food and Agriculture Organization/World Health Organization of the United Nations, Organization for Economic Cooperation and Development, EU Commission, French Academy of Sciences, National Research Council of the National Academy of Sciences, Royal Society of London and Society of Toxicology).

In contrast to traditionally bred crops, a rigorous safety-testing paradigm has been developed and implemented for GE crops, which utilizes a systematic, stepwise and holistic safety assessment approach (Cockburn, 2002; Kok and Kuiper, 2003; König et al., 2004). The resultant science-based process focuses on a classical evaluation of the toxic potential of the introduced novel gene, its gene product, and the wholesomeness for human consumption of the GE crop. In addition, detailed consideration is given to the history and safe use of the parent crop as well as that of the gene donor(s). The overall safety evaluation is conducted using the process known as "substantial equivalence (SE)," a model that is entrenched in all international crop biotechnology guidelines (Kok and Kuiper, 2003; Codex, 2003). The SE paradigm provides the framework for a comparative approach to identify the similarities and differences between the GE product and an appropriate comparator that has a known history of safe use. By building a detailed profile on each step in the transformation process (from parent to new crop) and by thoroughly evaluating the significance, from a safety perspective, of any differences that may be detected between the GE crop and its comparator, a comprehensive matrix of information is constructed. This information is used to reach a conclusion about whether food or feed derived from the GE crop is as safe as food or feed derived from its traditional counterpart or the appropriate comparator.

One common misunderstanding of the GE plant testing process involves the meaning and role of substantial equivalence. Some biotechnology critics have claimed that regulatory authorities have deemed all GE plants to be ipso facto substantially equivalent to their conventional counterparts, thereby requiring no safety testing at all. But as we describe above, the substantial equivalence concept merely describes the way in which GE plants should be tested. A conclusion that a product is substantially equivalent can only be reached after it is rigorously compared to a non-GE counterpart for material differences. Yet, even when a material difference is detected and the GE product is found to be not substantially equivalent to its conventional comparator, the finding does not automatically mean that the product is unsafe. Additional testing that explores these differences more completely may ultimately determine that the differences have no bearing on the product's safety for consumers or the environment.

Using this approach in the evaluation of more than 90 GE crops that have been approved in the U.S., the conclusion has been reached that foods and feeds derived from GE crops are as safe and nutritious as those derived from traditional crops. The lack of any credible reports of adverse effects resulting from the production and consumption of GE crops grown on more than 235 million cumulative hectares over the last seven years supports these safety conclusions.

The U.S. National Research Council in "Genetically Modified Pest-Protected Plants: Science and Regulation" (NRC, 2000) determined that no difference exists between crops modified through modern molecular techniques and those modified by conventional breeding practices. The NRC emphasized that the authors were not aware of any evidence suggesting foods on the market today are unsafe to eat because of genetic modification. In fact, the scientific panel concluded that growing such crops could have environmental advantages over other crops.

In a 2003 position paper, the Society of Toxicology (SOT, 2003) corroborated this finding and noted that there is no reason to suppose that the process of food production through biotechnology leads to risks of a different nature than those already familiar to toxicologists or to risks generated by conventional breeding practices for plant, animal or microbial improvement. It is therefore important to recognize that it is the food product itself, rather than the process through which it is made, that should be the focus of attention in assessing safety.

Similarly an EU Commission Report (EU, 2001, 2008) that summarized biosafety research of 400 scientific teams from all parts of Europe conducted over 15 years stated that research on GE plants and derived products so far developed and marketed, following usual risk assessment procedures, has not shown any new risks to human health or the environment beyond the usual uncertainties of conventional plant breeding. Indeed, the use of more precise technology and the greater regulatory scrutiny probably make GE plants even safer than conventional plants and foods. More recently the Commission funded research from 130 research projects involving 500 independent research groups over 25 years, concluding that "There is, as of today, no scientific evidence associating GE with higher risks for the environment or for food and feed safety than conventional plants and organisms" (Europa Press Release, 2010). Recent transcriptomic and metabolomic (Baker, 2006; Catchpole, 2005; Ricroch et al., 2011) studies in wheat and potatoes respectively show greater variation within and between conventionally bred cultivars and even growth locations than between GE and parental variety, except for the intended change. In fact, the differences within the same line between different geographical sites were generally greater than differences between various control and test lines at the same site.

The development of new tools for "omic" analysis (e.g., transcriptomics, proteomics and metabolomics) has prompted their evaluation for the safety assessment of transgenic crops since these kinds of untargeted holistic analytical methods offer the possibility of a more comprehensive insight into gene expression, protein content and detailed composition (Chassy, 2010). Although "omic" technologies have proven to be a powerful research tools, the application of these methods offers no useful information to the safety assessor. Transcriptomics is of limited value since measuring changes in gene expression cannot be translated directly into an understanding of changes in risk, or in metabolic or phenotypic traits. There are associated methodological, data handling and bias problems associated with transcriptomics as well and there is a paucity of baseline data that can be used to establish normal ranges. At this point in time, and for the foreseeable future, it is better to directly measure the outcome of genotype x environment interactions instead of trying to predict potential side effects based on observed changes in gene expression.

Proteomic analysis suffers from a similar set of problems (Chassy, 2010). Only a few hundred of the thousands of potential proteins in a cell can be evaluated; few can be evaluated with quantitative precision. More importantly, as is the case with transcriptomic analysis, estimation at a semi-quantitative level of which proteins are present in a cell, organ, tissue or organism, is not predictive of the composition of the organism, the phenotype of the organism, or any changes in risk associated with consumption of the organism. Proteomic analysis may, however, be useful as a tool for the rapid targeted analysis of specific proteins or sets of proteins as, for example, in the evaluation of changes in the content of allergens in a plant known to cause allergy.

The application of metabolomic profiling is more attractive from a safety assessment perspective since it focuses on the composition of the plant food or feed that is actually consumed (Chassy, 2010). Metabolomic profiling can be used for sample discrimination and classification (e.g., for GE comparator analysis), or for biochemical and mechanistic studies in discovery research and development. These tools may someday be employed to understand which parts of larger biochemical networks respond to genetic modification. While it has been claimed that unbiased non-targeted and comprehensive assessments can be improved by metabolite analyses, the usefulness of metabolomics as a safety assessment tool suffers from the same set of limitations that apply to transcriptomics and proteomics. At present, metabolomics is not one technology, but a family of analytical techniques each of which is capable of detecting the presence of tens to hundreds of metabolites. The data are difficult to reproduce; they are not quantitative; and baseline data and normal ranges for metabolites are not available. Moreover, it is often the case that many of the compounds detected are of unknown structure.

It is worth noting with regard to the usefulness, or lack of usefulness, of omic profiling for safety assessment, that the currently employed targeted analytical paradigm focuses on the measurement of all known macro- and micronutrients, toxicants, anti-nutrients, and phytochemicals of special interest in each specific crop (Harrigan and Chassy, 2012). The analysis often accounts for 95 percent or more of the composition of a sample. There is no evidence that targeted analysis is inadequate in identifying changes in composition that could be of concern to the risk assessor. Furthermore, it has been found that cultural, environmental and geographic differences often lead to large differences in composition between test samples of the same variety and these differences are often greater than those observed between different varieties of the same crop. Recall that the composition of transgenic crops more closely resembles that of their parental varieties than does that of other varieties of the same crop to the parental strain and to one another (Ricroch et al., 2011). At this point in time, omic profiling is not necessary to establish comparative safety, nor are standardized and validated omic methods ready for application in safety assessment.

The main principles of the international consensus approach to safety assessment of transgenic crops are listed below (Chassy et al., 2004; Chassy et al., 2008; Ricroch et al., 2011). They serve to illustrate the variety of principles that have been at the center of the discussions and that are continuously being updated:

- **Substantial equivalence:** This is the guiding principle for safety assessment. In short, substantial equivalence involves the process of comparing the GE product to a conventional counterpart with a history of safe use. Such a comparison commonly includes agronomic performance, phenotype, expression of transgenes and composition (macro- and micronutrients), and it identifies the similarities and differences between the GE product and the conventional counterpart. Based on the differences identified, further investigations may be carried out to assess the safety of these differences. These assessments include any protein(s) that are produced from the inserted DNA.

- **Potential gene transfer:** Where there is a possibility that selective advantage may be given to an undesirable trait from a food safety perspective, this should be assessed. An example is in the highly unlikely event of a gene coding for a plant made pharmaceutical being transferred to corn. Where there is a possibility that the introduced gene(s) may be transferred to other crops, the potential environmental impact of the introduced gene and any conferred trait must be assessed.

- **Potential allergenicity:** Since most food allergens are proteins, the potential allergenicity of newly expressed proteins in food must be considered. A decision-tree approach introduced by ILSI/IFBC in 1996 has become internationally acknowledged and recently updated by Codex FAO/WHO, 2003. The starting point for this approach is the known allergenic properties of the source organism for the genes. Other recurrent items in this approach are structural similarities between the introduced protein and allergenic proteins, digestibility of the newly introduced protein(s), and eventually (if needed), sera-binding tests with either the introduced protein or the biotechnology-derived product.

- **Potential toxicity:** Some proteins are known to be toxic, such as enterotoxins from pathogenic bacteria and lectins from plants. Commonly employed tests for toxicity include bioinformatic comparisons of amino acid sequences of any newly expressed protein(s) with the amino acid sequences of known toxins with those of introduced proteins, as well as rodent toxicity tests with acute administration of the proteins. In addition to purified proteins, whole grain from GE crops has been tested in animals, commonly in subchronic (90-day) rodent studies.

- **Unintended effects:** Besides the intended effects of the modification, interactions of the inserted DNA sequence with the plant genome are possible sources of unintended effects. Another source might be the introduced trait unexpectedly altering plant metabolism. Unintended effects can be both predicted and unpredicted. For example, variations in intermediates and endpoints in metabolic pathways that are the subject of modification while undesirable are predictable, while switching on/expression of unknown endogenous genes through random insertion in regions of the genome that control gene expression is both unintended and unpredictable. The process of product development that selects a single commercial product from hundreds to thousands of initial transformation events eliminates the vast majority of situations that might have resulted in unintended changes. The selected commercial product candidate event undergoes additional detailed phenotypic, agronomic, morphological and compositional analyses to further screen for such effects.

- **Long-term effects:** It is acknowledged that the premarket safety assessment should be rigorous to exclude potentially adverse effects of consumption of foods or feeds derived from GM crops. Nevertheless, some have insisted that such foods should also be monitored for long-term effects by postmarket surveillance. No international consensus exists as to whether such surveillance studies are technically possible without a testable hypothesis in order to provide meaningful information regarding safety, and a GM crop with a testable safety concern would most likely not pass regulatory review. The notion of using measurable biomarkers has been suggested but then these need to be determined for all foods and feeds whatever the source and balanced against reasonable economic burden.

The question of whether foods derived from organisms modified with recombinant DNA techniques should be specially labeled has received a great deal of attention. The FDA's (1992) approach to the labeling of foods, including those genetically engineered or otherwise novel, is that the label must be accurate and "material." Agency officials recognized that any breeding method could impart a change that makes food less safe or nutritious than its conventional counterpart, but that the process of rDNA modification is not inherently risky. Accordingly, special labeling is required "if a food derived from a new plant variety differs from its traditional counterpart such that the common or usual name no longer applies, or if a safety or usage issue exists to which consumers must be alerted."

Such changes include the introduction of a toxin, antinutrient or allergen into a food product in which consumers would not ordinarily expect to find it, such as an allergenic protein from nuts in corn; the elevation of an endogenous substance to potentially harmful levels, such as a significant increase in potato or tomato glycoalkaloids; or a significant change in the level of dietary nutrients in a food, such as oranges with abnormally low levels of vitamin C or a significant change in the lipid composition of cooking oils. Other material changes that must be labeled include those that relate to the storage, preparation or usage characteristics of a food, such as a change affecting the length of time or manner in which kidney beans must be soaked and cooked before eating or the safe shelf-life of various food products. Even a change in organoleptic characteristics of a food from what consumers would normally expect, including the taste, smell or mouth feel of a food, is considered material and must be labeled. Importantly, the FDA's policy stipulates that the altered characteristic itself must be specified on the label, not the breeding method used to impart the change. In that regard, labeling serves to alert consumers to the new characteristic of foods, not the processes used in their development.

The FDA's policy statement also emphasizes that no premarket review or approval is required unless characteristics of the biotech food explicitly raise safety issues, and that – in as much as the genetic method used in the development of a new plant variety does not meet either of the two criteria for "materiality" – the FDA cannot require the labeling to include this information. Obviously, many of the novel nutritionally enhanced foods expected on the market in the next few years will be labeled, as they will differ from their traditional counterparts, and in most instances, the company marketing them will want to proclaim their enhanced nutritional value.

4 Applications overview

Modifications of crop plants can be organized into two broad, non-mutually exclusive categories: those that benefit the producer through introduction of properties such as improved insect, weed and disease management and lower input costs, and those that benefit the consumer more directly with increased nutritional value, flavor or other desirable functional product attributes. Many plants in both categories also, either directly or indirectly, deliver benefits for the environment, such as reducing insecticide use and hastening an ongoing shift to conservation tillage practices. Modifications that protect the crop from either biotic or abiotic stress or that increase total crop yield primarily benefit the producer and are often called "input traits." (Biotic stress is damage by predators, such as insects and nematodes; weeds; or disease agents, such as viruses, fungi, and bacteria. Abiotic stress is damage from other, usually physical or climatic, causes, such as drought, flooding, cold, heat, salination and poor soils.) Scientists have just begun to tap the potential of biotechnology to produce varieties of plants that confer advantages to consumers directly. Varieties modified to have greater appeal to consumers are said to have enhanced "output traits." The majority of crops produced using modern biotechnology in present day commercial use fit in the former group.

In the United States, which has the largest number of approved and commercially planted biotech varieties, the U.S. Department of Agriculture has approved (in agency parlance, "deregulated") more than 90 transformation events of 16 plant species for commercial-scale cultivation (see http://www.aphis.usda.gov/biotechnology/petitions_table_pending.shtml) – though many of these products, while legal to grow and sell, are not commercially available. U.S. farmers also grow the largest number of acres (over 150 million) with biotech varieties, followed by Brazil (67m), Argentina (53m), India (23m), and Canada (23m). But by 2012, 28 countries had approved at least one transformation event for commercial cultivation, and these crops were grown by more than 17 million farmers on over 420 million acres total (James, 2013). Twenty of the 28 countries are emerging economies, and a full 90 percent of the 17.3 million farmers are resource-poor farmers from LDCs. Another 31 countries have permitted pre-commercial field trials with biotech crop varieties, or have approved some harvested biotech plants to be imported for use as food and livestock feed.

Among the varieties that are currently marketed, the most common traits are insect resistance, herbicide tolerance and virus resistance. And the most widely adopted species are the commodities corn, cotton, soy and canola. U.S. farmers grew each of these crops and planted a significant number of acres with biotech varieties of sugar beet and alfalfa, while a far smaller number of acres were planted with biotech squash, papaya and rice.

Among the most prevalent first generation products of agricultural biotechnology have been crop varieties resistant to certain chewing insects. This pest-resistance trait was added by inserting a gene from the common soil bacterium *Bacillus thuringiensis* (Bt), which produces an insoluble crystalline protein that adheres to and degrades the alkaline stomach of insect larvae, causing death. Different strains of the Bt bacterium produce proteins that are toxic to a specific range of insects, but not to mammals, fish, birds or other animals, including humans (EPA 2001). The bacterial proteins occur naturally, and organic farmers and foresters have cultivated Bt spores as a "natural pesticide" for decades. Due to the confinement of Bt protein's toxicity solely to insects – and a quite limited range of insects at that – it was a natural target for investigation by genetic engineers who hoped to use it in the production of insect resistant crops. By 2012, more than two dozen transformation events of corn, cotton, and potato with the Bt protein trait were "deregulated," or approved for commercial cultivation in the United States (USDA APHIS, 2012).

Experiments have been conducted with other genetic sources of insect resistance, such as lectin proteins (*Galanthus nivalis* agglutinin) (Gatehouse et al., 1999; Rao et al., 1998). And transgenic maize engineered for production of the medically useful protein avadin has been shown to produce effective resistance to a range of storage insect pests (Kramer et al., 2000). However, none of these traits has advanced beyond proof of concept stage for use in commercial food crop production.

Genetic engineering has been used to introduce tolerance to a number of different herbicides –including glyphosate, glufosinate, bromoxynil, imidazolinone, and others – which aids in effective weed management. Weeds compete with crop plants for sunlight, water, and soil nutrients, and if not eliminated, they can lead to significant yield losses. Consequently, effective weed management is essential to production-scale agriculture. Certain plant species are not damaged by many herbicides: Monocots, for example, generally are not harmed by selective herbicides designed to kill broad-leaf dicot species, and vice versa. Unfortunately, these selective herbicides are, by their very nature, unable to eliminate all weed species. Using rDNA methods, breeders have been able to insert genes that degrade specific herbicides or otherwise inhibit their action. Modified plants remain susceptible to other herbicides, however, so it is important for the seed and herbicide to be matched appropriately. In this way, growers can spray a broad-spectrum herbicide on their fields, effectively managing all or most weed species, while leaving the crop plants unharmed.

Herbicide tolerance can be introduced into crop species by several other breeding methods, including selection and induced mutation breeding. However, the genetic and biochemical mechanisms through which GE herbicide tolerant plants resist herbicides are well characterized and predictable, whereas breeders do not fully understand the genetic changes that take place in conventional plant breeding. So, while breeders continue to make use of conventional methods – one major seed company has in recent years used mutation breeding to introduce herbicide tolerance into an entire line of crop species, ranging from wheat to rice to canola – rDNA methods are generally more efficient and effective.

The last major class of biotech traits now on the market is virus resistance, and biotech resistant varieties of squash, papaya, plum and potato have been approved for commercial planting. There are now several ways of modifying plants to be resistant to viruses (Prins, Laimer, Noris, Schubert, Wassenegger, & Tepfer, 2008), but the resistance trait in all commercially approved GE plants arises from a phenomenon known as viral coat protein mediated resistance. Viruses are enclosed by a coat or "capsid" consisting of non-infectious proteins. Research conducted by Roger Beachy and his team at Washington University (Powell-Abel P., Nelson R.S., De B., Hoffmann N., Rogers S.G., Fraley R.T., & Beachy R.N., 1986) revealed that transforming a plant to express a gene that codes for a virus's capsid protein makes the plant resistant to the virus. The precise molecular mechanisms that govern coat protein mediated resistance are not fully understood (Bendahmane et al., 2007), but it is known that insertion of the viral gene does not infect the host plant or enable the plant to make new infectious particles. In fact, transgenic virus resistant plants express little if any of the coat protein itself, and consumers who eat such plants are often exposed to far higher levels of plant virus particles from infected plants in the human food supply (e.g., cauliflower mosaic virus). In neither case are consumers put at any risk, however, because plant viruses are not known to be pathogenic to humans and they have a long history of safe consumption.

Although critics have claimed that biotech crop production is helpful primarily to large farmers in industrialized countries, more than half of the 63 nations engaged in biotech research, development and production are less developed countries (LDCs). North America still leads in global production, with U.S. acreage accounting for about 43 percent of the total acreage worldwide. But 19 of the 29 countries that permit commercial-scale cultivation are LDCs. And some 15 million, or roughly 90 percent, of the farmers growing these crops are resource-poor farmers in those countries (James, 2012). The most recent countries to join this group include Uruguay, Paraguay, Bolivia, Egypt and Burkina Faso in 2009, followed by Pakistan and Myanmar in 2010. That year, Germany also resumed the legal planting of biotech crops (after withdrawing authorization several years earlier) by approving a potato variety modified to produce amylopectin starch for use as a polymer in industrial applications. Sweden, the first Scandinavian country to plant a biotech crop, also approved the high amylase starch potato. Though this product was not the first to incorporate a value-added output trait, some observers believe it signals a shift from the first generation of biotech crops focused largely on agronomic input traits, to the next generation, which will include far more output traits.

Developing and commercializing biotech plants with new or improved traits involves overcoming a variety of technical, regulatory and perception challenges inherent in perceived and real challenges of complex modifications. Where they have become available, many farmers have eagerly planted biotech crop varieties. And, by some estimates, biotech crops have been the most rapidly adopted agricultural technology in history – the very first commercial approval was granted in 1993, and biotech crops are now grown on roughly 10 percent of global cropland (James C., 2013). But the cultivation of biotech crops is still not authorized in most countries. And those that do permit commercial planting, or the use of harvested crops in food or animal feed, all subject biotech varieties to rigorous regulatory controls. Because of the added cost of meeting special regulatory requirements, the potential for lengthy delays in development and commercialization while awaiting regulatory approvals, and a host of public perception and consumer acceptance challenges, biotech crop innovation may not be as rapid as some advocates had hoped. Nevertheless, in order to meet the world's growing nutritional needs without substantially increasing agriculture's environmental footprint, both the panoply of traditional plant breeding tools and modern biotechnology-based techniques are likely to be necessary to produce plants with the desired agronomic and quality traits.

Furthermore, the tools of modern biotechnology will themselves continue to evolve, permitting breeders to modify plants in innovative new ways. In addition to the older gene transfer technologies, with which whole, single genes were introduced into plants, newer techniques, such as the use of RNA interference to manipulate endogenous gene expression and especially the use of transcription factors to modulate whole suites of genes and metabolic networks, will become increasingly important tools in the effort to introduce valuable traits. The later approach is already a major focus in the alteration of multigenic and quantitative traits such as developing stress tolerance crops and modifying metabolism for improving nutritional characteristics.

5 Agronomic traits and sustainability

Agricultural biotechnology has helped farmers around the world boost their productivity and grow crops in more ecologically healthy fields while allowing much more efficient use of resources. To give just one example, herbicide tolerant varieties now commercially available have accelerated an ongoing shift to conservation tillage (minimize cultivation) practices that decrease soil erosion as well as water and agricultural chemical runoff, and reduce machinery use, in turn decreasing fuel and water use and greenhouse gas emissions. Insect resistant varieties have reduced the need to spray synthetic chemical pesticides, with a corresponding reduction in farm worker pesticide poisonings and a lower impact on biodiversity.

Research by Brookes and Barfoot shows that, in the first fifteen years of GM crop cultivation, pesticide use (which includes both insecticides and herbicides) fell by over 448 million kg worldwide, or a decline of 9 percent (http://www.pgeconomics.co.uk/pdf/2012globalimpactstudyfinal.pdf 2012). In addition to the obvious direct benefits, less spraying means fewer tractor passes, contributing to lower carbon dioxide emissions. Meanwhile, the economic benefits experienced by large-scale and small-scale farmers in both industrialized nations and lesser-developed countries have been significant. The Brookes and Barfoot (2013) analysis shows that there have been very significant net economic benefits at the farm level amounting to $19.8 billion in 2011 and $98.2 billion for the 16 year period (in nominal terms). The majority (51.2%) of these gains went to farmers in developing countries. GE technology has also made important contributions to increasing global production levels of the four main crops, having added 110 million metric tons and 195 million metric tons, respectively, to the global production of soybeans and maize since the introduction of the technology in the mid-1990s. In addition, the environmental footprint associated with pesticide use was reduced by 15.4%, and there was a reduction in carbon dioxide emissions in 2010 equivalent to taking nearly 10 million cars off the road for a year (Brookes and Barfoot, 2012).

Though some critics have claimed that herbicide tolerant (HT) GE plants do nothing but encourage the use of agricultural chemicals, nothing could be farther from the truth. In addition to the soil health benefits discussed above, GE HT crops have several other environmental advantages. A study by researchers at Denmark's National Environmental Research Institute (NERI) monitored fields of conventional and glyphosate-tolerant sugar beet. They found that the GE plots supported more plant species and insects than the conventional plots, thus providing more food for birds and other types of wildlife. Thus, use of transgenic crops increased biodiversity compared to traditional herbicide treatments. The result is not surprising. One of the most important benefits arising from adoption of glyphosate tolerant plants is the resulting shift away from less environmentally benign herbicides. Compared with the herbicides it is replacing, glyphosate degrades more easily in soil and water. It has a negligible toxicity to mammals, birds, fish, frogs, crustaceans, bees, beetles and many soil organisms such as earthworms – although it does show moderate toxicity to some beneficial insects (Hin, Schenkelaars, & Pak, 2001).

Research comparing ploughed organic fields and no-till fields on the same UK farm, tended by the same farmer, indicate that no-till practices use only one-third as much fossil fuel, use land much more efficiently, reduce nitrate and pesticide run off by at least half and increase soil carbon, which is lost when ploughed. In addition, bird territories are orders of magnitude higher, soil erosion almost vanishes, and soil invertebrates such as earthworms/significantly increase in numbers, as do predatory arthropods that help keep pests down.

Although used by farmers the world over, cultivation with ploughs is not a sustainable practice. It is energy intensive, and it exposes soil to wind and water erosion. It allows rain to compact the soil and increases soil's oxygen content, thereby allowing organic matter to oxidize away. In turn, lower organic matter in the soil allows more compaction and more nutrient loss. Additionally in warmer and drier climates, tillage increases evaporative water loss, whereas conservation tillage results in lower evaporation as residue remains on the soil surface, creating a wetter and cooler soil microclimate. In short, reduced-till agriculture leads to healthier soil and reduced erosion.

Perhaps just as important, long-term tillage depletes soil carbon stocks by releasing it into the atmosphere in the form of CO2, methane and other greenhouse gasses. Conservation tillage practices on the other hand lead to far less carbon dioxide release. Soil carbon sequestration will be an important part of any international strategy to mitigate the increase in atmospheric CO2 concentrations. By adopting more sustainable management practices, agriculture can play a large part in enhancing soil carbon sequestration across the globe. To be sure, conservation tillage practices can be implemented by farmers that do not plant herbicide tolerant crop varieties, or by those who plant *non*-GE herbicide tolerant crops. Over the last few decades, conservation tillage has become a major component of the environmental stewardship practiced by farmers. However, the practice is far from universal. And problems associated with post-planting weed control were long ago identified as one of the biggest reasons why some farmers had not adopted conservation tillage (Koskinen and McWhorter, 1986). The widespread introduction of glyphosate tolerant crop plants in the U.S. in the mid-1990s, on the other hand, contributed to a significant acceleration of conservation tillage adoption.

The second major class of commercialized biotech crops, insect resistant Bt varieties, have also delivered considerable environment benefits. Brookes and Barfoot found that, from 1996 to 2010, use of GE insect resistant corn varieties resulted in the application of 42.9 million fewer kilograms of insecticide active ingredient globally, for nearly a 42 percent reduction. GE insect resistant cotton lowered insecticide application by 170.5 million kg, or a 24 percent reduction (Brookes and Barfoote, 2012).

Insect resistant maize also has a beneficial collateral effect: less insect damage results in much less infection by fungal molds, which in turn reduces mycotoxins that are known health risks causing such problems as liver cancer to humans and animals. The planting of Bt corn resulted in a 90% reduction in mycotoxin fungal fumonisins (Wild and Gong, 2010). In addition to the potential health benefits, the total U.S. economic benefit is estimated to be approximately $23 million annually. The only "natural" way to control those fungi is the use of copper sulfate, which has one of the highest toxic hazard ratings of acceptable pesticides and selects for antibiotic resistant bacteria in the soil. Researchers have also found that Bt crops' effective control of the European corn borer also results in a "halo effect" that helps suppress pest populations in neighboring non-GE fields (i.e, creates a cordon-sanitaire around the GE planted fields) (Hutchison et al., 2010; Wu et al., 2008, Carrière et al., 2003). European corn borer moths move between fields of Bt and non-Bt corn, and female borers lay eggs on both. Bt plants kill a substantial percentage of the hatched caterpillars, thereby greatly reducing regional pest populations with a beneficial impact on non-GE plants. In fact, Hutchison et al. (2010) found that non-GE corn neighboring Bt corn fields reaped greater economic benefit than did the biotech varieties because growers of the non-GE varieties enjoyed significant pest reductions without having to pay higher prices for seed.

While North America remains the epicenter for cutting edge GE research, other regions, namely China are emerging as contenders on the global stage. Agricultural science is now China's fastest-growing research field with China's share of global publications in agricultural science growing from 1.5 percent in 1999 to 5 percent in 2008. China's early experience with Bt cotton demonstrated the direct and indirect benefits of its investment in plant biotechnology research and product development. In 2002, Bt cotton was grown in 2.1 million hectares by around 5 million farmers. At that time the average Bt cotton farmer had reduced pesticide sprayings for the Asian bollworm from 20 to six times per year applications were reduced by 59-80 percent compared to conventional cotton (assessed in three years of use) and produces a kilogram of cotton for 28 percent less cost than the farmer using non-Bt varieties. Net revenues increased by $357-549/hectare compared to conventional cotton (assessed in three years of use).

Arguably more important are the social benefits from reducing exposure to insecticides and the impact of that on farm worker health. Bt cotton was first planted in China in 1997. Just three years later, Pray, Ma, Huang and Qiao found that farmers in northern China growing Bt cotton experienced over 75 percent fewer pesticide poisonings than those growing only non-GE cotton varieties (2001). Similarly dramatic reductions in pesticide poisonings were later reported from a broader study of Chinese cotton farmers (Huang et al., 2004), and among cotton farmers in South Africa (Bennett et al., 2003) and India (Kouser and Qaim, 2011).

The demand for productivity-enhancing technologies by farmers and for cost savings by consumers, the rate of increase in research investments and success with Bt cotton suggest that products from China's research program will one day become widespread inside that country. Indeed, China is emerging as one of the trend-setters in the adoption of novel traits as, more recently, China has begun to set the pace for new approvals, becoming the first major rice producing country to approve a GE rice variety in 2009, for example, and granting initial approval for a maize variety engineered to produce phytase, an enzyme that will reduce the amount of phosphate in the excrement/waste from corn-fed livestock (Anon., "China May Take Three Years to Allow Sales of Gene-modified Corn," BusinessWeek, 24 Feb. 2010). Rice is the principal staple for much of the world, and maize is the largest animal feed source, so rising productivity in those two crops will have important impacts on long-term food security. In pre-commercial field trials, Huang et al. (2005) found that Bt rice increased yields between 3.5 percent and 9 percent and decreased pesticide use by nearly 80 percent (17 kg/ha). The phytase maize approval is also a major step forward because it appears to be the only trait to date receiving its first approval in a less developed country. However, it is noteworthy in other ways, as this single trait addresses several issues, from nutritional to environmental, discussed further below.

The first GE crop to be released for commercial cultivation in India was Bt cotton, approved by the government in 2002 for field sowing in six states. In the first season of commercial cultivation, an estimated 44,500 hectares of certified Bt cotton were planted by nearly 55,000 farmers. The overall performance was mixed, as those initial cultivars thrived in regions for which they were originally developed but performed poorly in other areas with different climates and other challenges. It was not until the trait was introgressed by conventional breeding into locally adapted cultivars that Bt cotton thrived in all six growing regions.

From three Bt cotton hybrids in 2002 to 62 in 2006, the rapid deployment of Bt cotton hybrids specifically developed for different agro-climatic conditions resulted in a 39 percent reduction in insecticide sprays and 31 percent higher yields, resulting in increased profit per hectare of 88% or $250. During this same period, raw cotton exports rose from 0.9 million bales in 2005 to 4.7 million 2006 and 5.9 by 2007. By 2009 5.6 million resource-poor farmers in India planted 8.4 million hectares of Bt cotton, equivalent to 87 percent of the national cotton crop. The increase from 50,000 hectares when Bt cotton was first commercialized in 2002 to 8.4 million hectares in 2009 represents an unprecedented 168-fold increase in eight years. Between 2002 and 2008, Bt cotton generated economic benefits for farmers valued at $5.1 billion, halved insecticide requirements, contributed to the doubling of yield and transformed India from a cotton importer to a major exporter. Choudhary contends that the deployment of Bt cotton over the last eight years has resulted in India becoming the number one exporter of cotton globally as well as the second largest cotton producer in the world (Choudhary B. and Gaur K., 2008).

Despite the success of Bt cotton in India, the country's expected commercialization of Bt eggplant has not materialized because an effective opposition movement managed to block its approval. Field trials of Bt eggplant, or brinjal as it is referred to in India, was found to be effective against fruit and shoot borer (FSB), with 98 percent insect mortality in shoots and 100 percent in fruits compared to less than 30 percent mortality in non-Bt counterparts. Multi-location research trials confirmed that Bt brinjal required, on average, 77 percent less insecticide than non-Bt counterparts for control of FSB, and 42 percent less for the control of all insect pests of brinjal. The benefits of Bt brinjal translate to an average increase of 116 percent in marketable fruits over conventional hybrids, and 166 percent increase over popular open-pollinated varieties. Furthermore, the significant decrease in insecticide use reduces farmers' exposure to insecticides and results in a substantial decline in pesticide residues on brinjal fruits (Choudhary B. and Gaur K., 2010). The latter is quite significant since brinjal may be sprayed as many as 30 to 40 times in a growing season.

Scientists have estimated that Bt brinjal could deliver farmers a net economic benefit ranging from $330 to $397 per acre, with national benefits to India exceeding $400 million per year (Choudhary B. and Gaur K., 2010). However in February 2010, the environmental minister announced what was initially portrayed as a six-month moratorium on commercialization. Environment Minister Jairam Ramesh told reporters that additional testing was necessary to ensure the product was safe for consumers and the environment. However, he added that "public sentiment is negative" and it was his responsibility both to be attentive to science and "responsive to society" (see http://news.bbc.co.uk/2/hi/south_asia/8506047.stm). Ramesh explained that, "until we arrive at a political, scientific and societal consensus, this moratorium will remain," and the moratorium was still in place at the time of publication.

Similar stories of politicized decision-making can be found in dozens of other less developed countries, ranging from Asia to Africa to South America. For reasons as varied as fear of new technology and precautionary thinking to a concern for trade and the export market for foodstuffs, scores of countries have forbidden the planting of all GE crop varieties (Paarlberg 2001, 2010). And several countries, such as India, that do permit the cultivation of Bt cotton, permit no other crops – in large part because cotton is primarily a fiber crop, and only a small amount of cottonseed oil or meal are used for food and animal feed purposes. Of the 29 countries that have approved at least one GE crop for commercial cultivation, five currently permit only Bt cotton. Several others permitted only Bt cotton until recently. Nevertheless, 24 countries do currently permit the cultivation of GE crops for food and animal feed, most of them growing Bt maize or herbicide tolerant varieties of soy or canola (ISAAA 2012).

A second generation and beyond

The vast majority of products approved to date are in the area of input or agronomic traits, most specifically traits that counter biotic stress from insect pests, weeds, fungi and plant diseases. The principal focus in the immediate future will remain on expanding this first generation of agronomic traits, especially the area of pest control and stacking such traits. However, there is an increasing interest in abiotic stress tolerance, such as resistance to drought, extremes of heat or cold, and poor soil quality, which are gaining prominence as external pressures from a number of sources ranging from climate change to changing land use patterns and beyond.

Whereas the first generation of biotech crops generally incorporated a single novel gene, seed companies have increasingly begun to introduce varieties with multiple traits. On the biotic stress tolerance side the focus is expanding to multi-tiered control systems, such as two or more Bt genes with similar but not identical range of effectiveness, or a combination of insect resistance and herbicide tolerance genes. This, in theory, serves two purposes: Multiple Bt genes, for example, can expand the effectiveness of the insect resistance and, by exposing insects in the field to more than one Bt sub-species, may help to prolong the effectiveness of each resistance trait because there is less selective pressure from any one of them. On such example is SmartStax, an eight trait event developed through collaboration between Monsanto and Dow AgroSciences, which takes advantage of multiple modes of insect and nematode protection as well as tolerance to two different herbicides. Multi-tiered control systems like this are quickly becoming the norm.

Traits effective at addressing abiotic stresses have proven more difficult to develop, but several promising research programs are investigating a broad range of such traits, and a few are likely to reach the market within the next several years. And while the benefits of biotic stress tolerance in already commercialized crops has been substantial, successful abiotic stress tolerance traits may prove to have even greater benefits. A significant meta-issue overlays attempts to develop crops that are more resilient to abiotic stresses. Changing climatic conditions pose real challenges to available agricultural land and fresh water availability. Apart from its direct effects – higher temperatures, changing rainfall patterns, shifts in ecosystems, ocean acidification, and rising ocean levels – climate change is likely to have broader, more systemic effects that compound our current food security challenges. And all of these effects should be considered in the light of growing population levels, which could magnify food insecurity even further.

Severe drought, for example, accounts for half the world's food emergencies annually. In 2003, the World Food Program spent $565 million in response to drought in Sub-Saharan Africa (SSA). In this context solutions must be developed to adapt crops to, not only existing but also evolving, environmental conditions, such as marginal soils and greater extremes of heat, cold, drought, flooding and salinity.

The agriculture sector is both a contributor and provider of potential solutions to this phenomenon. Practices such as deforestation, methane from cattle flatulence and fertilizer use currently account for about 25% of greenhouse gas emissions. In total, agriculture contributes roughly fourteen percent of carbon dioxide emissions, forty eight percent of methane, and fifty two percent of nitrous oxide emissions. In addition, this sector uses a significant amount of available fresh water, supplies of which are likely to be further stressed by climate change. The FAO estimates that approximately seventy percent of the water currently consumed by humans is used in agriculture, and this is likely to increase as temperatures rise.

On the other hand, changes in agricultural technologies can play a substantial part in mitigating against climate change. This is especially relevant in emerging countries where producers and consumers are more subject to the mercy of the vagaries of climate fluctuations than in the west where there is greater capability of responding to the effects and managing resources. "Green biotechnology" offers a set of tools, which can help producers limit greenhouse gas emissions as well as adapt their agricultural techniques to shifting climates. The three major contributions of green biotechnology to the mitigation of the impact of climate change are greenhouse gas reduction, crop adaptation (environmental stress, changing niches) and protection and yield increase in less desirable and marginal soils.

On the first of these issues, greenhouse gas reduction, in addition to carbon dioxide, agriculture contributes two of the other major gases. Indeed, one of them, nitrous oxide, has a global warming potential of about 300 times that of carbon dioxide. In addition, nitrous oxides stay in the atmosphere for a considerable period of time. Nitrous oxide is produced through bacterial degradation of applied nitrogen fertilizer. Fertilizer can also contribute to eutrophication at ground level, so its reduction is desirable on several levels. However, nitrogen is essential for crop production since it is quantitatively the most essential nutrient for plants and a major factor limiting crop productivity. One of the critical steps limiting the efficient use of nitrogen is the ability of plants to acquire it from applied fertilizer. Therefore, the development of crop plants that absorb and use nitrogen more efficiently can serve both the plant and the environment.

Arcadia Biosciences of Davis, CA, developed nitrogen-efficient crops by introducing a barley AlaAT (alanine aminotransferase) into both rice and canola. Arcadia's Nitrogen Use Efficiency (NUE) technology produces plants with yields that are equivalent to conventional varieties but which require significantly less nitrogen fertilizer because the AlaAT gene allows more efficient use (Arcadia NUE, 2013). Compared with controls, transgenic plants also demonstrated significant changes in key metabolites and total nitrogen content, confirming increased nitrogen uptake efficiency. This technology has the potential to reduce the amount of nitrogen fertilizer that is lost by farmers every year due to losses to air, soil and waterways. In addition to environmental pressures, nitrogen costs can represent a significant portion of a farmer's input costs and can significantly impact farmer profitability. Farmers spend $60 billion annually for 150 million tons of fertilizer. The technology has been licensed to Dupont for maize and to Monsanto for application in canola.

The second area where green technology can help in a changing climate is crop adaptation to environmental stress and changing niches. Under stress, plants will divert energy into survival instead of producing biomass and reproduction, so addressing this impact should have substantial effect on yield. In addition, improved stress tolerance allows expanded growing season, especially earlier planting, and it further reduces yield variability and grower financial risk.

Arguably the most critical abiotic stress is lack of sufficient water. Yet one of the most effective methods of addressing water limitation problems, namely irrigation, unfortunately is also one of the major causes of arable land degradation, as mineral salts that occur naturally in irrigation water accumulate over time in soils. It is estimated that 24.7 million acres of farmland worldwide is lost each year due to salinity build up resulting from over irrigation. In fact, crops are now limited by salinity on 40 percent of the world's irrigated land (including 25 percent of the U.S.). To address this salinity problem, Eduardo Blumwald at UC Davis has developed plants with the AtNHX1 gene from *Arabidopsis thaliana* to make them better tolerate soil and water salinity. In *A. thaliana*, AtNHX1 mediates the transport of Na+ and K+ into the plant vacuole. By overexpressing this vacuolar Na+/H+ antiporter, transgeneic tomatoes were able to grow, flower and produce fruit when irrigated with water containing 200 mM of sodium chloride, roughly half the salinity of ocean water (Sottosanto et al, 2007). Arcadia Biosciences has now introduced this gene into economically important crops.

The most critical time for water stress is near pollination and flowering, when yields with or without irrigation can vary by up to 100 percent. This effect is clearly demonstrable in dry land production, where yields can be cut in half in the absence of irrigation. At this time, about 15 percent of U.S. maize acres are irrigated, but it is estimated that about 70-80 million acres in the U.S. suffer yield losses due to moderate water stress. Given the negative effects of water stress and the cost of irrigation, it is estimated 20 million acres in U.S. would benefit from a drought tolerance gene that affords a ten percent yield increase over non-irrigated crops. It would also allow shifting of higher value crops into production on more marginal land.

One of the first commercialized products to have included a "yield gene" is Monsanto's second generation Roundup Ready 2 Yield Soybeans (RR2Y), which include not only the glyphosate tolerant trait but which was developed using extensive gene mapping to identify specific DNA regions that segregated with yield increase. Marker assisted selection was used to introgress the desired trait. It is a perfect example of the power of combining recombinant DNA technology with genomics tools. The company claims that following four years of field trials across six U.S. states RR2Y showed seven to eleven percent higher yields, compared to the first generation of Roundup Ready soybeans.

As noted previously, the introduction of specific transcription factors (Tfs) are versatile tools being employed in the development of stress tolerant plants. One of the most versatile classes of transcription factors, in so far as environmental response is concerned, is the DREB (dehydration-responsive element binding protein) transcription factors. These proteins are involved in the biotic stress signaling pathway and can activate as many as twelve additional genes that trigger plant resistances/or tolerance to such stresses as drought, freezing and salt. It has been possible to engineer stress tolerance in transgenic plants by manipulating the expression of DREBs. One Tf gene isolated from Arabidopsis has improved drought tolerance increasing productivity by at least two-fold during severe water stress (Gosal, 2009). In Monsanto field trials using this approach, maize yields have increased under water stress by up to 30 percent compared to controls.

Other approaches include modification of individual genes involved in stress response and cell signaling. For example, drought tolerant canola engineered to reduce the levels of PARP (poly[ADP-ribose] polymerase), a key stress-related protein in many organisms, show relative yield increases of up to +44 percent compared to control varieties. A subset of the transcription factors homeodomain leucine zipper proteins (HDZip) play a role in regulating adaptation responses including developmental adjustment to environmental cues such as water stress in plants. One of these effectors is abscisic acid (ABA), an important plant regulator controlling many environmental responses including stomata opening which is itself modulated by the DREB elements. Some work is being done on modifying HDZip directly and other are working indirectly such as for example down regulating farnesyltransferase, a signaling system in the production of abscisic acid and stomata control, which results in stomata closure and water retention.

UC Davis's Eduardo Blumwald is also working on modifying abscisic acid levels to enhance the tolerance of plants to water deficit by delaying the drought-induced leaf senescence and abscission during the stress episode. Using tobacco plants expressing an isopentenyltransferase (IPT) gene under the control of a stress- and maturation-induced promoter (PSARK), Blumwald's team showed that delayed drought-induced leaf senescence resulted in remarkable drought-tolerant phenotypes, as well as minimal yield loss when plants were watered with only 30 percent of the water used under controlled conditions (Zhang, 2010). This trait is now being introduced into rice among other crops. This work is being done in conjunction with Arcadia Biosciences. In addition, Bayer CropScience, Pioneer Hi-Bred, BASF and Dow among others are conducting research on maize, cotton, canola and rice, to develop a new generation of stress-tolerant, high-performance crop varieties. Clearly, stress tolerant traits are of paramount importance in LDCs, especially those in Sub-Saharan Africa and Asia. The partnership, known as Water Efficient Maize for Africa (WEMA), was formed in response to a growing call by African farmers, leaders and scientists to address the devastating effects of drought on small-scale farmers. Frequent drought leads to crop failure, hunger and poverty. Climate change can only aggravate this situation.

On the other end of the spectrum of climate change impact is flooding due to changing rain patterns and rising sea levels. This is already a major cause of rice crop loss. It is estimated that 4 million tons of rice are lost every year because of flooding which is sufficient to feed 30 million people. Rice is not grown in flooded fields through necessity but rather to control weeds, however, most rice varieties die after more than three days of complete submergence. Researchers knew that at least one rice variety, FR13A, can tolerate flooding for longer periods, but conventional breeding failed to create an event that was acceptable to farmers. The Ronald laboratory at UC Davis cloned the submergence tolerance (Sub1) locus from this resistance variety using a map-based cloning approach. The Sub1 locus encodes three putative transcription regulators one of which increases dramatically in response to oxygen deprivation in sub1 seedlings; whereas Sub1C levels decrease. Transgenic lines that over-express the Sub1A-1 gene have been introgressed into a submergence intolerant line and the progeny displayed enhanced submergence tolerance. This has also been achieved using Marker assisted selection.

There is also some research in the final abiotic stress focus area namely expansion of crops into, and increased yield in, less desirable and marginal soils. For example, a gene that produces citric acid in roots can protect plants from soils contaminated with aluminum as it binds to the contaminant preventing uptake by the root system. In naturally acidic or alkaline soils, which comprise over 70 percent of the world's arable land, phosphorous forms compounds with elemental aluminum and other metals, making the phosphorous unavailable for plant absorption. Adding phosphorous fertilizer to the soil helps plants grow to full maturity. But because large amounts of added phosphorous goes un-used by plants, run-off becomes a significant pollution problem. Scientists at the Center for Research and Advanced Studies in Irapuato, Mexico, have bioengineered tobacco and papaya plants with a gene from the bacterium *Pseudomonas aeruginosa* to secrete citric acid from their roots (Lopez-Bucio et al., 2000; de la Fuente et al., 1997). Genes such as these can allow crops to be cultivated in hostile soils and temperatures, thereby increasing geographic range while reducing negative impacts on waterways and fragile ecosystems.

6 Renewable resources

With the increasing costs, in both economic and environmental terms, of our dependency on fossil fuels, biotechnology offers innovative means to improve plant material and processing enzymes for biomass conversion into fuels. There are two principal classes of biofuels bio-alcohol, initially bio-ethanol but with increasing interest in higher energy alcohols such as bio-butanol; and bio-diesel (Knothe, 2007).

The first generation of biofuels was fermented from readily available and easily accessible sources of simple sugars, such as sugarcane and simple polysaccharides – primarily from grain starch. These sources of bioethanol are unsustainable on many levels, including the fact that they compete with food and feed grains for markets, land and water. The focus for second generation bio-alcohols is mostly on complex polymers that are not used for food or grain, primarily cellulosic ethanols and what are being termed third generation bio-alcohols such as bio-butanol. From a biotechnology perspective, research is being conducted on the biomass component focusing on increased production in such sources as switchgrass and miscanthus by, among other things, modifying photoperiodicity genes to switch energy to vegetative tissue production and improved biomass conversion by such approaches as reducing lignin composition and incorporating self-activating enzyme digestion upon harvesting.

Bioprocessing enzyme producers including Novozymes (Davis, CA) Novozymes biofuels (Novozymes, 2013) and Danisco (Palo Alto, CA) are making considerable strides in improving the effectiveness and specificity and lowering the cost of cellulosic enzymes as well as increasing the conversion range especially for the more difficult pentose sugars such as xylose. Protein engineers, on the other hand, are taking a synthetic biology approach. Recently, progress has been reported on engineering more stable and effective enzymes such as cellobiohydrolases by researchers at Caltech and completely novel synthetic metabolic pathways by the Berkeley company Amyris (Amyris, 2013).

Biodiesel is a mono-alkyl ester, typically made via a trans-esterification process reacting lipids (vegetable oil, animal fat) with an alcohol. One advantage of biodiesel over other biofuels is that it can be used directly by standard diesel engines, whereas most biofuels and waste oils can only be used to fuel specially converted diesel engines. Biodiesel can be used alone, or blended with petro-diesel. It has better lubricating properties and much higher cetane ratings than today's lower sulfur diesel fuels. But is still not yet economical as an alternate stand-alone fuel. Promising research has now begun on using modified algae to produce biodiesel fuel in a much more economically and environmentally sustainable way than either cellulosic or other land-based sources. From a biotechnology perspective, the main focus for expanding interest in this area is increasing lipid production and modifying lipid composition for optimum performance. Work is being done to modify algae for increased production of desirable medium chain fatty acids (MCFA), which eliminates the requirement for cracking and isomerization of long chain fatty acids (LCFAs). The advantages of MCFAs over LCFAs are higher energy density, lower fuel viscosity, lower flash point, and lower freezing point. (Knothe, 2007).

While researchers have had success in producing biodiesel and other biofuels from algae in small pilot projects, these have not yet been successfully scaled up to commercial-scale production. To date, contamination, de-wetting, and lipid isolation remain economically prohibitive. Nevertheless, the Department of Energy estimates that algae fuel may eventually yield as much as 30 times more energy per acre than land crops such as soybeans (Scott, 2010).

7 Post-harvest characteristics

The FAO estimates that between 10 and 40 percent of global food is lost post-harvest through poor transportation systems, poor storage conditions and inadequate warehousing, insect damage, fungal damage, loss to pest animals and waste in commercial and domestic processing (FAO, 1989). It is not unusual in many parts of the world for high quality agricultural products to ripen correctly and subsequently spoil prior to consumption. But there are several ways in which biotechnology can contribute to reduced post-harvest loss of foods.

The first genetically engineered crop product approved for sale was a tomato in which ripening had been modified through the introduction of antisense genes. Calgene, Inc., of Davis, California, commercialized the FLAVR SAVR tomato, modified through antisense RNA to reduce production of the enzyme polygalacturonase. Polygalacturonase degrades the "glue" that holds plant cell walls together and, for this reason, is the central enzyme in softening (and increased susceptibility to disease) of ripe tomatoes. A tomato with reduced production of this enzyme can be picked red and more flavorful, rather than green, and still survive transport to market. The gene that encodes the polygalacturonase enzyme is transcribed into another nucleic acid called messenger RNA (mRNA). This is then translated into the protein/enzyme. When Calgene placed the gene coding for the polygalacturonase enzyme back into the tomato genome in the reverse orientation, the two mRNAs annealed and effectively blocked translation of the enzyme (Bruening and Lyons, 2000). From a scientific perspective, the GE tomato worked well. However, for a variety of reasons, including high initial cost and public concerns about biotech foods, the FLAVR SAVR tomato proved unsuccessful commercially. Nevertheless, the product provided an important proof of concept leading to additional research into ways of reducing post-harvest spoilage, with some additional successes. For example, efforts also have been directed at using antisense technology to switch off one of the genes involved in the production of ethylene, the key hormone in fruit ripening. The pathway for the biosynthesis of ethylene is well understood.

The enzyme ACC synthase converts SAM (S-adenosylmethionine) to ACC (1 aminocyclopropane-1-carboxylic acid), which is in turn converted to ethylene by ACC oxidase. Florigene has isolated carnation genes for both ACC synthase and ACC oxidase. By using antisense technology to suppress expression of these genes ethylene production is blocked, leading to flowers that have a long life after cutting. The same result has also been achieved by co-suppression. Introducing additional copies of the genes for ethylene synthesis leads to suppression of both the transgene and the native genes. This research is being extended to other crops including broccoli, raspberries and bananas.

Another strategy for increasing shelf-life will be more widely applicable to leafy crops and involves regulating the level of cytokinin, the hormone responsible for delaying leaf senescence. The Agrobacterium gene ipt encodes an enzyme isopentenyl transferase which catalyses a key step in cytokinin biosynthesis. Researchers in the laboratory of Richard Amasino at the University of Wisconsin placed the ipt gene under control of the promoter from a senescence-associated gene isolated from *Arabidopsis*. They were able to demonstrate that, in transgenic tobacco, the leaf cytokinin level is auto-regulated – that is, as the leaf ages, the promoter is activated, triggering just enough cytokinin production to reverse the senescence process, without disturbing other aspects of plant development (Gan, 1997). A spectacular delay in the ageing (yellowing) of the leaves was observed. McCabe (2001) achieved the same in lettuce.

This observation clearly has exciting implications for the shelf-life of vegetable crops. Normally, leaves yellow rapidly after harvest as the supply of cytokinin from the roots is severed. A European Consortium, co-ordinated from Ireland, has now demonstrated similarly spectacular results to Amasino's in lettuce, cauliflower and tomato. In fruit crops like tomato, the benefits could lie in extending the photosynthetic life of the plant, which might positively affect yield while also reducing fungal infection (because fungi favour senescent leaves).

Nutrition

At the most fundamental level, food is viewed as a source of nutrition to meet minimum daily calorific requirements for survival, but with increasing prosperity there is an ever greater focus on the desire for food that promotes optimal health. Aside from the basic nutrition perspective, there is a clear dichotomy in demonstrated need between different regions and socioeconomic groups, the starkest being injudicious over-consumption in the industrialized world leading to obesity and under-nourishment starvation in less developed countries (LDCs). Both extremes suffer from forms of malnourishment – one through inadequate supply, the other, in many but not all instances, through inappropriate choices. Dramatic increases in the occurrence of obesity, cardiovascular disease, diabetes, cancer and related ailments in developed countries are in sharp contrast to chronic under- and genuine malnutrition in many LDCs. Although, unfortunately, populations in many LDCs are beginning to experience both under-nutrition, in terms of micronutrient deficiency, and negative health impacts arising from over-consumption of high-calorie, low-nutrient grains and other staples.

Both problems could be addressed to some degree by a modified food supply. And the tools of biotechnology, while not offering a singular solution, can play a significant role in ameliorating global malnutrition. Worldwide, plant-based products comprise the vast majority of human food intake, irrespective of location or financial status (Mathers, 2006). In some cultures, either by design or default, plant-based nutrition comprises almost 100 percent of the diet. Given this fact, one can deduce that significant nutritional improvement can be achieved via modifications of staple crops to deliver higher micronutrient levels.

While the correlative link between food and health, beyond meeting basic nutrition requirements, has only been unequivocally proven in a number of cases, a growing body of evidence indicates that food components can influence physiological processes at all stages of life. Nutrition intervention from a functionality perspective has a personal dimension. Parsing individual response is at least as complex a challenge as the task of increasing or decreasing the amount of a specific protein, fatty acid, or other component of the plant itself. There is also evidence that early food regimes in childhood can affect later life health. For example, some children that survived famine conditions in certain regions of Africa grew into adults who battled obesity and obesity-related problems such as diabetes, possibly due to the selective advantage of the so-called "thrifty gene" in their early food-stressed environment. According to the thrifty gene hypothesis, a genetic adaptation in early human evolution would have helped hunter-gatherer populations store fat more readily during times of abundance, enabling thrifty gene carriers to better survive periods of food scarcity. In today's era of relative food abundance, presence of the thrifty genotype becomes a hazard in the presence of calorie dense diets.

Functional foods are defined as any modified food or food ingredient that may provide a health benefit beyond the traditional nutrients it contains. Scientific evidence is accumulating to support the role of phytochemicals and functional foods in the prevention and treatment of disease. Functional food components are of increasing interest in the prevention and/or treatment of a number of the leading causes of death including cancer, diabetes, cardiovascular disease, and hypertension (diseases of the developed world). Many food components are known to influence the expression of both structural genes and transcription factors in humans. Examples of these phytochemicals are listed in Table 2. The large diversity of phytochemicals suggests that the potential impact of phytochemicals and functional foods on human and animal health merits their examination as potential targets of biotechnology research.

Table 2 - Examples of Plant Components with Suggested Functionality[a]

Class/ Components	Source[b]	Potential Health Benefit
Carotenoids		
Alpha-carotene	Carrots	Neutralizes free radicals that may cause damage to cells.
Beta-carotene	Various fruits, vegetables	Neutralizes free radicals.
Lutein	Green vegetables	Contributes to maintenance of healthy vision
Lycopene	Tomatoes and tomato products (ketchup, sauces)	May reduce risk of prostate cancer.
Zeaxanthin	Eggs, citrus, maize	Contributes to maintenance of healthy vision.
Dietary Fiber		
Insoluble fiber	Wheat bran	May reduce risk of breast and/or colon cancer.
Beta glucan[c]	Oats	May reduce risk of cardiovascular disease (CVD).
Soluble fiber[c]	Psyllium	May reduce risk of CVD.
Whole Grains[c]	Cereal grains	May reduce risk of CVD.
Collagen Hydrolysate	Gelatin	May help improve some symptoms associated with osteoarthritis
Fatty Acids		
Omega-3 fatty acids - DHA/EPA	Tuna; fish and marine oils	May reduce risk of CVD and improve mental, visual functions.
Conjugated Linoleic acid (CLA)	Cheese, meat products	May improve body composition, may decrease risk of certain cancers.

[a] Examples are not an all-inclusive list

[b] U.S. Food and Drug Administration approved health claim established for component.
Modified from Newell-McGloughlin, 2008.

Class/ Components	Source[b]	Potential Health Benefit
Gamma Linolenic Acid	Borage, evening primrose	May reduce inflammation risk of cancer, CVD disease and improve body composition.
Flavonoids		
Anthocyanidins: cyanidin	Berries	Neutralize free radicals, may reduce risk of cancer.
Hydroxycinnamates	Wheat	Antioxidant-like activities, may reduce risk of degenerative diseases.
Flavanols: Catechins, Tannins	Tea (green, catechins), (black, tannins)	Neutralize free radicals, may reduce risk of cancer.
Flavanones	Citrus	Neutralize free radicals, may reduce risk of cancer.
Flavones: quercetin	Fruits/vegetables	Neutralize free radicals, may reduce risk of cancer.
Glucosinolates, Indoles, Isothiocyanates		
Sulphoraphane	Cruciferous vegetables (broccoli, kale), horseradish	Neutralizes free radicals, may reduce risk of cancer.
Phenolics		
Stilbenes – Resveratrol,	Grapes	May reduce risk of degenerative diseases; heart disease; cancer. May have longevity effect.
Caffeic acid, Ferulic acid	Fruits, vegetables, citrus	Antioxidant-like activities; may reduce risk of degenerative diseases; heart disease, eye disease.
Epicatechin	Cacao	Antioxidant-like activities; may reduce risk of degenerative diseases; heart disease.
Plant Stanols/Sterols		
Stanol/Sterol ester[c]	Maize, soy, wheat, wood oils	May reduce risk of coronary heart disease (CHD) by lowering blood cholesterol levels.
Prebiotic/Probiotics		
Fructans, Inulins, Fructo-oligosaccharides (FOS)	Jerusalem artichokes, shallots, onion powder	May improve gastrointestinal health.

Class/ Components	Source[b]	Potential Health Benefit
Lactobacillus	Yogurt, other dairy	May improve gastrointestinal health.
Saponins	Soybeans, soy foods, soy protein-containing foods	May lower LDL cholesterol; contains anti-cancer enzymes.
Soybean Protein	Soybeans and soy-based foods	25 g/day may reduce risk of heat disease.
Phytoestrogens		
Isoflavones - Daidzein, Genistein	Soybeans and soy-based foods	May reduce menopause symptoms, such as hot flashes, reduce osteoporosis, CVD.
Lignans	Flax, rye, vegetables	May protect against heart disease and some cancers; may lower LDL cholesterol, total cholesterol, and triglycerides.
Sulfides/Thiols		
Diallyl sulfide	Onions, garlic, olives, leeks, scallions	May lower LDL cholesterol, helps to maintain healthy immune system.
Allyl methyl trisulfide, Dithiolthiones	Cruciferous vegetables	May lower LDL cholesterol, helps to maintain healthy immune system.
Tannins		
Proanthocyanidins	Cranberries, cranberry products, cocoa, chocolate, black tea	May improve urinary tract health. May reduce risk of CVD, and high blood pressure

From a health perspective, plant components of dietary interest can be broadly divided into four main categories, which can be further broken down into positive and negative attributions for human nutrition.

- macronutrients (proteins, carbohydrates, lipids [oils], and fiber),
- micronutrients (vitamins, minerals, phytochemicals),
- anti-nutrients (substances such as phytate that limit bioavailability of nutrients),
- allergens, intolerances and toxins

Macronutrients: Protein

The FAO estimates that about one billion people worldwide suffer from under-nutrition, of which insufficient protein in the diet is a significant contributing factor (FAO, 2012). Protein-energy malnutrition (PEM) is the most lethal form of malnutrition and affects every fourth child worldwide (WHO, 2006). Most plants have a poor balance of essential amino acids relative to the needs of animals and humans. The cereals (maize, wheat, rice etc.) tend to be low in lysine, whereas legumes (soybean, peas) are often deficient in the sulfur-rich amino acids, methionine and cysteine. Successful examples of improving amino acid balance to date include high-lysine maize (Eggeling, Oberle, & Sahm, 1998; O'Quinn et al., 2000) canola and soybeans (Falco et al., 1995). Free lysine is significantly increased in high lysine maize by the introduction of the dapA gene (cordapA) from *Corynebacterium glutamicum* that encodes a form of dihydrodipicolinate synthase (cDHDPS) that is insensitive to lysine feedback inhibition. Consumption of foods made from these crops potentially can help to prevent malnutrition in developing countries, especially among children.

Another method of modifying storage protein composition is to introduce heterologous or homologous genes that code for proteins containing elevated levels of the desired amino acid such as sulfur containing (methionine, cysteine) or lysine. An interesting solution to this to create a completely artificial protein containing the optimum number of the essential amino acids methionine, threonine, lysine, and leucine in a stable, helical conformation designed to resist proteases to prevent degradation. This was achieved by a number of investigators, including sweet potato modified with an artificial storage protein (ASP-1) gene (Prakash CS, 2000). These transgenic plants exhibited a two- and five-fold increase in the total protein content in leaves and roots, respectively, over that of control plants. A significant increase in the level of essential amino acids such as methionine, threonine, tryptophan, isoleucine, and lysine was also observed (Chassy et al., 2008; Prakash CS, 2000). A key issue is to ensure that the total amount and composition of storage proteins is not altered to the detriment of the development of the crop plant when attempting to improve amino acid ratios (Rapp, 2002).

Some novel indirect approaches have also been taken to improve protein content. An ancestral wheat allele that encodes a transcription factor (NAM-B1) that accelerates senescence and increases nutrient remobilization from leaves to developing grains (modern wheat varieties carry a nonfunctional allele) has been "rescued" (Uauy et al., 2006). Reduction in RNA levels of the multiple NAM homologs by RNA interference delayed senescence by more than three weeks and reduced wheat grain protein, zinc, and iron content by more than 30 percent. Yet another approach to indirectly increase protein and oil content has been used (Young et al., 2004). They used a bacterial cytokinin-synthesizing isopentenyl transferase (IPT) enzyme, under the control of a self-limiting senescence-inducible promoter, to block the loss of the lower floret resulting in the production of just one kernel composed of a fused endosperm with two viable embryos. The presence of two embryos in a normal-sized kernel leads to displacement of endosperm growth, resulting in kernels with an increased ratio of embryo to endosperm content. The end result is maize with more protein and oil and less carbohydrate (Chassy et al., 2008).

Macronutrients: Fiber and Carbohydrates

Fiber is a group of substances chemically similar to carbohydrates that non-ruminant animals including humans poorly metabolize for energy or other nutritional uses. Fiber provides bulk in the diet such that foods rich in fiber offer satiety without contributing significant calories. There is ample scientific evidence to show that prolonged intake of dietary fiber has various positive health benefits, especially the potential for reduced risk of colon and other types of cancer.

When colonic bacteria (especially *Bifidobacteria*) ferment dietary fiber or other unabsorbed carbohydrates, the products are short-chain saturated fatty acids. These may enhance absorption of minerals such as iron, calcium, and zinc, induce apoptosis preventing colon cancer and inhibit 3-hydroxy-3-methylglutaryl coenzyme-A reductase (HMG-CoAR) thus lowering low density lipoprotein (LDL) production (German et al., 2005). Dietary fructans and inulins, among other polysaccharides, have been shown to promote the growth and prevalence of gastrointestinal bifidobacteria in humans. Supplementation of the diet with so-called "pro-biotic" poly-fructans has been proposed as means to promote the growth of gastrointestinal "probiotic" bifidobacteria which should be health-beneficial. Plants are effective at making both polymeric carbohydrates (e.g., starches and fructans) and individual sugars (e.g., sucrose and fructose). The biosynthesis of these compounds is sufficiently understood to allow the bioengineering of their properties and to engineer crops to produce polysaccharides not normally present. Polymeric carbohydrates such as fructans have been produced in sugar beet and inulins and amylase (resistant starch) in potato (Hellwege et al., 2000) without adverse effects on growth or phenotype. A similar approach is being used to derive soybean varieties that contain some oligofructan components that selectively increase the population of beneficial species of bacteria in the intestines of humans and certain animals and inhibit growth of harmful ones (Bouhnik et al., 1999).

Macronutrients: Novel Lipids

As noted previously, genomics and marker assisted plant breeding combined with recombinant DNA technology, provide powerful means for modifying the composition of oilseeds to improve their nutritional value and supply the functional properties required for various food oil applications. Genetic modification of oilseed crops can offer an abundant, relatively inexpensive source of dietary fatty acids with potentially wide ranging health benefits. Production of such lipids in vegetable oil provides a convenient mechanism to deliver healthier products to consumers without the requirement for significant dietary changes. Major alterations in the proportions of individual fatty acids have been achieved in a range of oilseeds using conventional selection, induced mutation and, more recently, post-transcriptional gene silencing. Examples of such modified oils include: low- and zero-saturated fat soybean and canola oils, canola oil containing medium chain fatty acids (MCFA) whose ergogenic potential may have application in LDCs, high stearic acid canola oil (for trans fatty acid-free products), high oleic acid (monounsaturated) soybean oil and canola oil containing the polyunsaturated fatty acids (PUFA), gamma-linolenic (GLA; 18:3 n-6) and soybean oils containing stearidonic acids (SDA; C18:4 n-3) (Stearidonic acid, 2011) very-long-chain fatty acids (Zou et al., 1997) and omega-three fatty acids (Yuan and Knauf, 1997; Stearidonic acid, 2011). These modified oils are being marketed and many countries have a regulatory system in place for the pre-market safety review of novel foods produced through conventional technology.

Edible oils rich in monounsaturated fatty acids provide improved oil stability, flavor and nutrition for human and animal consumption. High-oleic soybean oil is naturally more resistant to degradation by heat and oxidation, and so requires little or no post-refining processing (hydrogenation), depending on the intended vegetable oil application. Oleic acid (18:1), a monounsaturate, can provide more stability than the polyunsaturates, linoleic (18:2) and linolenic (18:3). Antisense inhibition of oleate desaturase expression in soybean resulted in oil that contained > 80 percent oleic acid (23 percent is normal) and had a significant decrease in PUFA (Kinney AJ, 1998). Dupont have introduced soybean oil composed of at least 80 percent oleic acid, and linolenic acid of about 3 percent, and over 20 percent less saturated fatty acids than commodity soybean oil. Monsanto's Vistive contains less than 3 percent linolenic acid, compared to 8 percent for traditional soybeans. These result in more stable soybean oil, and less need for hydrogenation. The genetically modified version Vistive gold (MON 87705) is engineered to reduce linolenic acid content by suppressing FATB and FAD2, endogenous enzymes that play a role in the biosynthesis of fatty acids. This alteration more than triples oleic acid content, raising it from approximately 20 to 70 percent of all fatty acids, and reduces the levels of linoleic acid, stearic acid and palmitic acid present in seeds (Vistive Gold Soybeans, 2011).

A key function of alpha-linolenic acid (ALA) is as a substrate for the synthesis of longer-chain omega-3 fatty acid found in fish, eicosapentaenoic acid (EPA; C20:5n-3) and docosahexaenoic acid (DHA; C22:6n-3), which play an important role in the regulation of inflammatory immune reactions and blood pressure, brain development in utero, and, in early postnatal life, the development of cognitive function. Stearidonic acid (SDA, C18:4n-3), EPA, and DHA also possess anti-cancer properties (Christensen, Christensen, Dyerberg, & Schmidt, 1999; Reiffel and McDonald, 2006; Smuts et al., 2003). Research indicates that the ratio of n-3 to n-6 fatty acids may be as important to health and nutrition as the absolute amounts present in the diet or in body tissues. Current Western diets tend to be relatively high in n-6 fatty acids and relatively low in n-3 fatty acids. Production of a readily available source of long-chain-PUFA, specifically w-3 fatty acids, delivered in widely consumed prepared foods could deliver much needed w-3-fatty acids to large sectors of the population with skewed n-6:n-3 ratios. In plants, the microsomal w-6 desaturase-catalyzed pathway is the primary route of production of polyunsaturated lipids. Ursin et al. (2000, 2003) has introduced the delta-6 desaturase gene from a fungus (Mortierella) succeeding in producing omega-3 in canola. In subsequent work, the same gene was added to soybean and transgenic soybean oil was obtained that contains > 23 percent SDA, with an overall n-6: n-3 ratio of 0.5., which the body converts to heart-healthy eicosapentaenoic acid (EPA), one of three omega-3 fatty acids used by the body. This product is now being developed for commercialization by Monsanto (Stearidonic Acid (SDA, 2011)

Micronutrients: Vitamins and minerals

Micronutrient malnutrition, the so-called hidden hunger, affects more than one-half of the world's population, especially women and preschool children in developing countries (UN SCN, 2004). Even mild levels of micronutrient malnutrition may damage cognitive development and lower disease resistance in children, and increase incidences of childbirth mortality. The costs of these deficiencies, in terms of diminished quality of life and lives lost, are large (Pfeiffer and McClafferty, 2007). The clinical and epidemiological evidence is clear that selected minerals (iron, calcium, selenium and iodine) and a limited number of vitamins (folate, vitamins E, B6 and A) play a significant role in maintenance of optimal health and are often limiting in diets, particularly where dietary diversity is low (e.g., among the poor).

While increasing dietary diversity is an effective way to reduce micronutrient and vitamin deficiencies, another strategy to ensure an adequate dietary intake of specific micronutrients is to adjust their levels in commonly consumed plant foods. Using various approaches including genomics, vitamin E levels are being increased in several crops, including soybean, maize and canola, while rice varieties are being developed with the enhanced vitamin A precursor, beta-carotene, to address vitamin A deficiency that leads to macular degeneration and impacts development. Golden Rice II accumulates up to 37 µg of beta-carotene per gram of rice (23-fold more than the original). This beta-carotene has been shown to be bioavailable in sufficient amounts that 100 to 200g per day can provide adequate provitamin A to ameliorate against deficiency (Tang et al., 2009). Golden Rice was finally approved for commercialization in the Philippines in February 2013. A number of other staple crops on which many depend almost exclusively for calories have been produced enriched in beta-carotene, including maize, cassava, millet and sorghum (Harjes et al., 2008; Welsch et al., 2010; Yan et al., 2010). Cassava is being field tested in Nigeria. Ameliorating another major deficiency in LDCs namely minerals such as iron and zinc has also been addressed. Iron is the most commonly deficient micronutrient in the human diet, and iron deficiency affects an estimated 1 to 2 billion people. Anemia, characterized by low hemoglobin, is the most widely recognized symptom of iron deficiency, but there are other serious problems such as impaired learning ability in children, increased susceptibility to infection, and reduced work capacity. Drakakaki et al. (2005) demonstrated endosperm-specific co-expression of recombinant soybean ferritin and Aspergillus niger phytase in maize which resulted in significant increases in the levels of bioavailable iron. A similar end was also achieved with lettuce (Goto et al., 2000). The Africa Biofortified Sorghum (ABS) Project developed the world's first sorghum transformation system as well as the first "golden" sorghum that had elevated pro-vitamin A levels, reduced phytate, a raised grain protein profile and the raised absorbability of zinc and iron (Blaine, 2011).

A rather interesting approach to increase the levels of calcium in crop plants was taken by (Connolly, 2008) by using a modified calcium/proton antiporter (known as short cation exchanger 1 [sCAX1]) to increase Ca^{2+} transport into vacuoles. They also demonstrated that consumption of such Ca^{2+}-fortified carrots results in enhanced Ca^{2+} absorption. Other targets include development of folate-enriched tomatoes and enhancement of antioxidant isoflavonoids (DellaPenna, 2007; Yonekura-Sakakibara et al., 2007).

Micronutrients: Phytochemicals

Approximately 25,000 metabolites of the more than 200,000 phytochemicals produced by plants are thought to have potential health-protective or health-beneficial value in the human diet. Analysis of these metabolites, most specifically metabolomic analysis, is a valuable tool in better understanding what has occurred during crop domestication through the identification of lost and silenced traits. This information can be applied to the design of new paradigms for more targeted crop improvement with respect to phytochemical content.

Decades of research in nutrition have demonstrated many of the specific mechanisms that make vitamins and minerals essential for human health. On the other hand, the primary evidence for the health-promoting roles of phytochemicals comes from epidemiological studies, and the exact chemical identity of many of the bioactive compounds present in various plant foods that appear to have an associated with improved health outcomes have yet to be determined. However, for a few plant-derived phytochemicals, such as non-provitamin A carotenoids, glucosinolates and phytoestrogens, the active compound or compounds have been identified and are being rigorously studied.

Epidemiological studies have suggested a potential benefit of the carotenoid lycopene in reducing the risk of prostate cancer, particularly for the more lethal forms of this cancer. Five studies point to a 30 percent to 40 percent reduction in risk associated with high tomato or lycopene consumption in the processed form in conjunction with lipid consumption, although other studies with raw tomatoes were not conclusive (Giovannucci, 2002). In a study by (Mehta et al., 2002) to modify polyamines to retard tomato ripening, an unanticipated enrichment in lycopene was found, with levels up by 2- to 3.5-fold compared to conventional tomatoes. This is a substantial enrichment, exceeding that so far achieved by conventional means, and this approach may work in other fruits and vegetables as well. Likewise, as noted previously, Gonzali et al. (2009) used snapdragon transcription factors to achieve high levels of the reactive oxygen scavengers, anthocyanins expression in tomatoes.

Other phytochemicals of interest include related polyphenolics such as resveratrol which has been demonstrated to inhibit platelet aggregation and eicosanoid synthesis in addition to protecting the sirtuins, genes implicated in DNA modification and life extension; flavonoids, such as tomatoes expressing chalcone isomerase that show increased contents of the flavanols rutin and kaempferol glycoside; glucosinolates and their related products such as indole-3 carbinol (I3C); catechin and catechol; isoflavones, such as genistein and daidzein; anthocyanins; and some phytoalexins (Table 1). A comprehensive list of phytochemicals is outlined in Table 2. As research demonstrates some or all of them to have special value in maintaining or improving consumer health, biotechnologists will no doubt explore the possibility of adding the genes that code for various phytochemicals to food plants, or increasing their expression levels where they already appear. In summary, although there is a growing knowledge base indicating that elevated intake of specific phytochemicals may reduce the risk of diseases, such as certain cancers, cardiovascular diseases and chronic degenerative diseases associated with aging, further research and epidemiological studies are still required to prove definitive relationships. Modern molecular methods may prove useful in investigation of the relationship between dietary intake of specific phytochemicals and human health outcomes, and in the future plants with enhanced content of specific phytochemicals may be developed by plant breeders.

Anti-nutrients, Allergens, Toxins

Plants possess many defense strategies that have evolved to protect them from predators. Some of these compounds, such as resveratrol and glucosinolates, which are primarily pathogen protective chemicals, also have been demonstrated to have beneficial effects on human and animal health. Many, however, can have the opposite effect. For example, many glucosinolates have both protective and anti-nutrient properties. And phytate, a plant phosphate storage compound, is considered an anti-nutrient as it strongly chelates the essential mineral micronutrients iron, calcium, zinc and other divalent mineral ions, making them unavailable for absorption by humans and certain animals. Non-ruminant animals generally lack the enzyme phytase which is needed for metabolic breakdown of phytate. Poultry and swine producers must add processed phosphate to their feed rations to ensure adequate levels are bioavailable in the diet. Phosphate that is not absorbed by the animals is excreted into the environment resulting in water pollution. When low-phytate soybean meal is utilized along with low-phytate maize for animal feeds phosphate excretion in swine and poultry manure is halved.

A number of groups have added heat-and acid-stable phytase from *Aspergillus fumigatus* or other organisms to make the phosphate and essential divalent metal ions more bioavailable in several crops (Potrykus, 1999). A gene for a metallothionein-like protein has also been introduced into plants to promote the reabsorption of iron. Low-phytate maize was commercialized in the USA in 1999 (Wehrspann, 1998). In November 2009, the Chinese company Origin Agritech announced the final approval of the world's first genetically modified phytase-expressing maize (Han, 2009). Research indicates that the protein content of low-phytate soybeans is also slightly more digestible than the protein in traditional soybeans. Better results were obtained using transgenic plant material in a poultry feeding trial than with the commercially produced phytase supplement (Keshavarz, 2003). Poultry grew well on the bioengineered alfalfa diet without any inorganic phosphorus supplement, which demonstrates that plants can be engineered to increase the bioavailability of this essential mineral.

Other antinutrients that are being examined as possible targets for reduction are trypsin inhibitors, lectins, and several heat-stable components found in soybeans and other crops. Likewise strategies are being implied to reduce or limit food allergens (albumins, globulins, etc.), malabsorption and food intolerances (gluten) and toxins (glycoalkaloids, cyanogenic glucosides, phytohemagglutinins) in crop plants and aesthetics undesirables such as caffeine (Ogita S, 2003). Examples include changing the levels of expression of the thioredoxin gene to reduce the intolerance effects of wheat and other cereals (Buchanan et al., 1997) and using RNAi to silence the major allergen in soybeans (P34 a member of the papain superfamily of cysteine proteases) and rice (14-16 kDa allergenic proteins). Blood serum tests indicate that p34-specific IgE antibodies could not be detected after consumption of gene-silenced beans (Helm et al., 2000; Herman et al., 2003).

Biotechnology approaches can be employed to down-regulate or even eliminate the genes involved in the metabolic pathways for the production, accumulation, and/or activation of toxins in plants. For example, the solanine content of potato has already been reduced substantially using an antisense RNA approach, and efforts are underway to reduce the level of the other major potato glycoalkaloid, chaconine (McCue KF, 2003). Work has also been done to reduce cyanogenic glycosides in cassava through expression of the cassava enzyme hydroxynitrile lyase in the roots (Siritunga & Sayre, 2003). When "disarming" plants' natural defenses in this way, one must be aware of potentially increased susceptibility to pests, diseases and other stressors. There are a number of possible strategies to deal with deletion of host defenses, but these are beyond the scope of the present paper.

Barriers to introduction

9

Although biotech crops are now grown by more than 16 million farmers in 29 countries, the technology has nevertheless met stiff resistance from some consumers, producers, NGOs activists and regulators in numerous countries. Many countries ban both the cultivation of GE crops and the import of food or animal feeds derived from them. Yet, even in the countries where GE crops are grown widely, such as the United States and Canada, the vast majority of production is limited to the commodity grains corn, canola and soy, which are primarily fed to livestock or consumed by humans only after processing, and cotton. Despite the significant economic benefit reaped by producers of GE commodity crops, very few GE varieties of whole fruits or vegetables are grown commercially. The explanation for this phenomenon is complex and multi-faceted, but consumer attitudes, food industry ambivalence, production costs, regulatory impediments and market access all play a role.

U.S. consumer attitudes tend to be mixed regarding food biotechnology, though most Americans know very little about the subject. In a 2012 International Food Information Council survey, just 38 percent of respondents said they had a favorable opinion of using biotechnology to produce food. On the other hand, only 20 percent had a negative opinion, while most were undecided or said they did not know enough to form an opinion. Still, 77 percent said they would be likely to purchase foods bioengineered to require less pesticide use, and 71 percent said they would buy foods made with cooking oils modified to have a healthier fat content. Because both such products already exist, this may seem to be evidence of significant consumer support for food biotechnology. However, the lack of widespread knowledge means that few consumers are aware of the benefits the accompany biotech products. Furthermore, a small but important segment of those who hold anti-biotechnology attitudes are very passionate about those beliefs. So, many packaged food companies and food retailers have been reluctant to openly embrace biotechnology for fear that there was more to lose than gain from doing so.

Public attitudes also vary broadly outside the United States, with consumers in many European countries expressing the most significant opposition. Still, as Kalaitzandonakes et al. (2005) have documented, publicly expressed opinions do not always predict shopper behavior. For example, over a period of several years in the late 1990s, numerous public opinion surveys found that large majorities of consumers in the Netherlands expressed significant concern about genetically engineered foods. Nevertheless, the authors found neither abrupt nor gradual shifts away from products made with GE ingredients after the Netherlands introduced mandatory GE food labeling in 1997. Nor did consumers significantly change their purchasing behavior three years later when the Dutch mandatory labeling law was revoked and consumers could no longer easily distinguish between GE and non-GE products.

Similarly, Hur (2001) reported that the appearance of labeled GE foods in one Japanese grocery store chain did not have a noticeable impact on consumer purchases. Runge and Jackson (2000) found that the appearance of milk labels touting products from cows not treated with the GE growth hormone rbST did not result in consumer defection from unlabeled products containing milk from treated cows. And Noussair et al. (2004) found in that 91 percent of the participants in their experimental auction study of French consumers initially said they were unwilling to purchase GE products, but that 42 percent later indicated they would be willing to purchase them if the products were sufficiently inexpensive. Arguably the most interesting finding in the Noussair et al. study is the fact that the overwhelming majority of participants did not actually examine package labels to determine whether products did or did not contain GE ingredients. The results of these and other studies suggest that, despite their stated preferences, most consumers do not in fact reject foods with GE ingredients at the point of sale. Nevertheless, retailer concerns about rejection by a small minority of consumers may contribute to an unwillingness to sell those products.

Packagers and retailers also report a reluctance to subject their valued brand reputations to attacks from anti-biotechnology activists (Miller and Conko, 2003). Perhaps the most telling example of retailer deselection is the experience with GE varieties of russet Burbank potatoes, first introduced in Canada and the United States in 1998 and 1999 respectively. The varieties, engineered with a Bt gene for resistance to the Colorado potato beetle and resistance to potato leaf role virus proved initially popular with growers. Within a year, however, several major fast food companies, including McDonald's and Burger King, became the targets of anti-biotechnology campaigners and soon pledged not to use GE potato varieties. The restaurant chains told their french-fry suppliers to stop using engineered potato varieties, and the products were removed from the market in 2001 (Kilman 2001). A much bigger problem than consumer resistance is the rejection of biotech foods by major producers and retailers. With anti-biotechnology campaigners so eager to protest against supermarket chains and food processing companies who use bioengineered ingredients, it is understandable that few firms are willing to put their hard-earned brand reputations at risk. And the bigger the companies, the less willing they seem to use biotechnology (Kalaizandonakes and Bijman, 2003).

Anecdotal evidence regarding consumer and retailer preferences aside, there have been only a small number of GE whole foods brought to market in the United States. To date, these include the FlavrSavr tomato mentioned above, varieties of virus-resistant squash and papaya, herbicide-tolerant and Bt insect-resistant varieties of sweet corn and a few others. Arguably the most significant contributing factor to this small trickle of products is the regulatory cost of testing and seeking approval for engineered varieties. Redenbaugh and McHughen (2004) report that, in the United States alone, regulatory compliance adds at least $1 million to the cost of developing a GE variety for each transformation event.

For crops that are shipped in international commerce, the regulatory costs associated with market approval in key producing and importing countries have been estimated to range from $6 million to $15 million (Kalaitzandonakes et al., 2007) up to as much as $35 million (McDougall, 2011). Of course, regulators in some important export markets where public anti-biotechnology sentiment is high will simply not approve GE varieties of certain crop species. Yet even if approval is eventually granted, importers and retailers in such countries may nevertheless reject such products due to the marketability concerns described above. U.S. wheat farmers, for example, have refused to plant a GE herbicide tolerant wheat variety approved for commercial planting in 2004 because it has not been approved for food use in Europe or Asia. And countries ranging from China to Argentina have refused to authorize certain GE crop varieties that cannot be exported to important overseas markets (Paarlberg, 2001).

Multinational corporate crop developers appear willing to bear such great costs for highly valuable commodity grains because global sales are large enough to justify the heightened regulatory expenditures. But researchers in the public sector, as well as those at charitable organizations and small startup companies, rarely have sufficient resources to navigate the complex and expensive regulatory approval process for any new varieties. Most of them must pass off their research to larger, better financed corporations to bring to market. But even these well-funded researchers often cannot justify the expense of developing GE varieties of lower-market value products, such as horticultural crops or the staple crops grown primarily by subsistence farmers in less developed countries. The result has been that researchers may conduct numerous, promising laboratory and greenhouse experiments with GE fruits and vegetables or such developing world staples as millet, sorghum, and cassava, but the products most likely to advance through field testing to regulatory approval and commercialization are of the major commodity species, including corn, cotton, soy and canola.

Although corporate-funded researchers in industrialized countries often have little difficulty affording the expense of complying with such regulations, few public sector researchers have the financial resources to devote to sometimes duplicative and often unnecessary laboratory or field assessments. Similarly, wealthy countries may be able to afford the significant opportunity costs that result from an unjustifiably burdensome regulatory apparatus. But such rules add little to environmental and human health protection, and arguably do far more harm than good in less developed countries.

Complicating matters further is the highly charged political environment in which regulatory decisions are made, and the near constant allegations that any approvals could prove catastrophic. Biosafety regulators, particularly those in LDCs where lack of confidence in regulatory capacity is often heightened, have an incentive to move slowly because they know that anti-technology campaigners and the media will challenge them whenever they release GE crops for commercial use. Presaging the attitude expressed by Indian authorities nearly a decade later, after approving Bt cotton varieties in 2002, the chairman of India's Genetic Engineering Advisory Committee told *Newsweek International*, "We took a lot of flak over GE cotton ... It was my job to ensure we weren't accused of over-hastiness" (Guterl, 2003). Thus, we find evidence that regulators make decisions not solely on products' merits, but on the basis of their own perceptions of public attitudes about the likelihood of regulators making mistakes.

Worldwide there is clear asymmetry and lack of consensus in regulatory systems. This discourages research on less economically valuable, low volume or otherwise specialized crops and traits and is a real disincentive to creative research. For all intents and purposes, there is just one trait from a public institution that has successfully traversed the regulatory minefields and been translated into a commercially viable commodity. A team of USDA Agricultural Research Service (ARS) researchers, led by Dennis Gonsalves at Cornell University, developed the viral coat protein protection system, initially used to combat the papya ringspot virus pandemic in Hawaii. The PRSV-resistant papaya, based on RNAi suppression of the coat protein expression, literally saved the $17 million Papaya economy in Hawaii and is significant in Taiwan and other Southeast Asian countries. After PRSV was discovered in the major papaya growing area of Hawaii in the early 1990s, plans were initially developed to cut down every tree in order to contain the virus's spread. Commercial introduction of PRSV-resistant papaya in 1997 obviated the need for such drastic measures, however, and the Hawaiian papaya industry is thriving today, with no evidence that the virus resistance is eroding (Gonsalves, 2010).

ARS researchers have subsequently used the same technology to develop plum pox resistant plum trees (Scorza et al., 2001). Plum pox is the most threatening viral disease of stone fruits, such as plums, peaches, nectarines, apricots and cherries, and once it appears in a given region it is nearly impossible to eradicate without burdensome quarantine requirements and destruction of infected trees. Fortunately, plum pox has been detected in the U.S. on only a small number of occasions and was successfully eradicated each time. But, like PRSV-resistant papaya, the C5 Honey Sweet plum variety could one day prevent a substantial collapse of the U.S. plum crop were the virus to gain a significant foothold on U.S. soil. One other GE variety developed by a public institution, herbicide tolerant flax developed by Alan McHughen's team at the University of Saskatchewan (McHughen and Holm, 1995), was approved for commercialization in Canada and the United States in 1998 and 1999 respectively. However, due to an inability to secure approval in the commercially important European export market, the product was never widely planted and has since been removed from the market altogether.

While translation of biotech research into value added products for producers and consumers is a challenge in the US it is exponentially more difficult in LDCs. One problem facing Africa in particular is the lack of a dynamic private sector to take technologies to the farmer. It has also been estimated that regulatory costs might exceed the costs of research and experimentation needed to develop a given GE crop, which is a major problem in releasing such crops to the market. A way to reduce the costs of generating food and environmental safety data is to develop regional "centers of excellence" with complementary facilities where food safety testing can be done reliably and regulatory costs could be reduced. Despite their potential productivity and environmental advantages, several LDC governments in Africa have explicitly indicated a reluctance to approve biotech crop varieties for fear of jeopardizing access to important export markets (Paarlberg, 2009).

The economic gains from using genetically modified crop technology in Sub-Saharan Africa (SSA) are potentially large according to the World Bank Group (Andersen, Jackson, & Nielson 2004: see http://www-wds.worldbank.org/external/default/WDSContentServer/IW3P/IB/2004/09/08/000009486_20040908122621/additional/101501322_20041117144001.pdf). The results suggest the welfare gains are potentially very large, especially from

Golden Rice and nutritionally enhanced GE wheat, and that those benefits are diminished only slightly by the presence of the European Union's current ban on imports of GE foods. The authors conclude that, if SSA countries impose bans on GE crop imports in deference to EU market demand for non-GE products, the domestic consumer loss net of that protectionism boost to SSA farmers would be more than the small gain derived from greater market access to the EU.

More practical problems cited for the slow passage of GE crops from experimental, to trial, to commercial stage in LDCs include the lack of capacity to negotiate licenses to use genes and research techniques patented by others, especially for crops with export potential. In addition, there are difficulties in meeting regulatory requirements and a lack of effective public commercialization modalities and working extension networks. Biosafety and IPR regulations still have to be enforced in many countries for an effective and safe use of genetically engineered crops, especially if their production is meant for the export market.

Intellectual property constraints also appear to be a significant barrier in some regions and for some technologies. At the research level this is usually not of major import in developing regions due to research exemptions in existing patent law. However, the presence of intellectual property rights can make it prohibitively expensive for public sector researchers and even financially constrained private sector researchers to move products from the laboratory or greenhouse to commercialization. In some cases, this may be because licenses to use individual patents are too expensive to obtain, while in others the problem may arise simply because the transaction costs of tracing and licensing multiple patents is beyond the technical capacity of non-specialists. Even when intellectual property owners – which may be either private companies or other public sector institutions – are willing to grant patent licenses at little or no cost, lack of familiarity with the process of searching for patents and negotiating licenses to use them can prove challenging. One key example of this phenomenon occurred with the development of Golden Rice, whose commercial development necessitated the negotiation of licenses or other contracts to use more than 40 protected genes or processes. Although the holders of these patents were willing to grant licenses for use of Golden Rice in LDCs, negotiating the agreements required a substantial effort and significant legal expertise (Kryder, Kowalski, & Krattinger, 2000, ISAAA Briefs No.20).

A few non-profit ventures have been developed in recent years to help public sector researchers navigate the complex maze of intellectual property rights. The Public Intellectual Property Resource for Agriculture (PIPRA) was established at the University of California, Davis, in 2004 to assist public sector agriculture researches better address intellectual-property and technology-transfer issues. PIPRA's initial goal was to create a user-friendly, searchable database of relevant patents in order to help public sector institutions better manage their patent portfolios and licensing obligations and to facilitate the commercialization of public sector agricultural innovations (Atkinson et al., 2003 http://www.sciencemag.org/content/301/5630/174.full). The organization now provides a much broader range of intellectual property and regulatory compliance services to public sector scientists. The U.S. Department of Agriculture's Economic Research Service also offers a searchable database of patented agricultural technologies to assist public sector researchers with intellectual property licensing and management. And the Center for Application of Molecular Biology to International Agriculture (CAMBIA) in Australia not only provides its own database of intellectual property, but is also building an "open source" platform of agricultural technologies through which researchers can use, share, and build upon innovations that are in the public domain (Dennis, 2004: http://www.nature.com/nature/journal/v431/n7008/full/431494a.html).

Coexistence

10

But what of the context in which these crops are grown? Can all cropping systems co-exist in harmony? According to Brookes and Barfoot (2008), it is important to determine the relative importance of different crop production systems based on planted area, production and economic value to the region in question. The issue is what, if any, are the economic consequences of adventitious presence of material from one crop system within another, based on the notion that farmers should be able to cultivate freely the crops of their choice using whichever production system works best in any given context (GE, conventional or organic). It is never a food or environmental safety issue but rather a production and marketing matter. The heart of the issue is assessing the likelihood of adventitious presence of material from one production system affecting another and the potential impacts. This requires consistency when dealing with the adventitious presence of any unwanted material including, but most definitely not limited to, biotech-derived material. Adventitious presence is simply the unintended incidence of something other than the desired crop such as small quantities of weed seeds, seeds from other crops, dirt, insects or foreign material (e.g., stones). It is unrealistic to expect 100 percent purity for any crops, or products derived there-from, so thresholds that are consistent across all materials should be set and should not discriminate (e.g., thresholds for adventitious presence of biotech material should be the same as applied to thresholds for other unwanted material and vice versa). All measures should be proportionate, non-discriminatory and science-based.

The issue of economic liability provisions that compensate growers for adventitious presence of biotech material is often raised and is hotly contested. Historically, legal systems have placed the burden of preventing adventitious presence of unwanted material in an agricultural crop on the grower of that crop. For example, seed breeders bear the legal burden of isolating certified seed crops from undesired pollination by a neighbor's crop of the same species. Breeders, not their neighbors, must use geographic isolation, physical barriers such as trees or hedgerows, and buffer zones to separate seed and non-seed crops. By extension, many biotechnology advocates have suggested, the onus of preventing unwanted cross pollination by biotech plants should fall on growers of any specialty crops, such as organic grains, fruits and vegetables.

Many organic growers and others who wish to preserve the non-biotech identity of their crops counter that biotechnology and biotech crops are sufficiently different from historical forms of "contamination" that the burden of preventing cross-pollination should fall on the growers of biotech varieties. EU coexistence rules place the burden on biotech growers.

Virtually all EU member states have transcribed EU Directive 2001/18 and implement EU regulations on traceability and labeling. Within the EU, provision has been made for a de minimis threshold for unavoidable presence of GE, but no actual threshold has been set. Therefore, the default state of the 0.9 percent on labeling and traceability is the one enforced.

Indeed, there have been at least two occasions to date in which the presence of minute levels of transgenes in non-biotech crops have given rise to legal concerns. The transgene from Bayer Crop Sciences's unapproved LLRICE 604 were found in Clearfield 131 (CL131) rice seed in 2007, and the transgene from Mycogen's unapproved event "32" was discovered in maize in 2008. U.S. regulatory agencies were able to move quickly and determine that these events did not prove any risk because they carried similar constructs to those already having achieved non-regulated status. The problem nevertheless made the "contaminated" rice and corn unable to be exported, causing financial losses of millions of dollars for growers. In those cases, involving as yet unapproved biotech varieties, U.S. courts determined that the developers did in fact have an obligation to ensure isolation from non-biotech and approved biotech crops.

For fully approved varieties, however, growers of specialty crops who have themselves chosen a more stringent standard than that established by the market for basic commodity products should not expect their neighbors to bear the special management costs of meeting that self-imposed standard. To do so would reverse fundamental freedoms of economic activity and would establish a dangerous precedent. It would allow specialty operators to formulate unrealistic standards for biotech production, would impose impossibly high standards on neighbors, and would effectively impose a ban on the choice of other producers. Specialty crop growers usually are rewarded by higher prices and niche markets for taking such actions. Their neighbors enjoy no such advantage. Furthermore, because organic certification is process based, not content based, accidental cross-pollination of an organic crop by biotech plants does not in fact affect the organic status of the resulting food or feed.

If new regulations were adopted to address economic liability provisions for any negative economic consequences of adventitious presence of unwanted material, one might argue that the same principle should apply to all farmers regardless of their chosen production methods. On equity grounds, biotech growers should have equal access to compensation for adventitious presence of material from conventional or organic crops (such as fungal contamination) as conventional and organic producers have from biotech growers. All co-existence measures should be based on legal, practical and scientific realities and not on commercial or niche marketing objectives.

According to Brookes and Barfoot (2008) fully approved biotech crops do generally co-exist successfully with conventional and organic crops in North America (where, as noted, biotech crops account for the majority of acreage of important arable crops like soybeans, cotton and maize) Spain and the more recently the Czech Republic. The market has developed practical, proportionate and workable coexistence measures without new regulations or indeed any government intervention. Where isolated instances of adventitious presence of biotech material have been found in conventional or organic crops these have usually been caused by inadequate implementation of good coexistence practices (e.g., inefficient segregation of crops in storage and transport, nonuse of tested certified seed).

Under civil liability (i.e., tort damages), and for intellectual property infringement (except for the unauthorized StarLink), there have been no lawsuits brought by any parties for adventitious presence. Every case brought by a seed company for infringement has involved a claim that the farmer charged with infringement was an intentional infringer (i.e., adventitious presence was not the issue). And, to date, each of these cases was upheld by the courts. Indeed, all except one notable exception in North America has conceded to this claim.

Using the American Seed Trade Association's recommended practices as the starting point, peaceful coexistence can be accomplished through voluntary cooperative practices that include the following four principles:

1. Monitoring: Verify the models and predictions about cost, isolation standards, and generally to learn how the farming community copes with the requirements for keeping the product streams separated.
2. Dialogue: Strategy development takes place in a dialogue between the scientific and technical community and all relevant stakeholders.
3. Stewardship: Stewardship programs should take into account the interests of both GE and non-GE farmers. Existing product stewardship programs for non-GE crops in farming should be a starting point for developing stewardship schemes for GE crops.
4. Research: The scientific community should be encouraged to fill the knowledge gaps that have been identified. Projects are needed to validate models and guidelines, including long-term studies. Much is already known about how identity preservation from more than a century of experience by seed breeders in their efforts to ensure the genetic purity of certified seed. However, additional research can improve mechanistic, probabilistic and predictive models of gene flow in an effort to beef up our ability to prevent unwanted cross pollination. In addition, scientists have already developed biological methods for restricting gene flow by eliminating the fertility of pollen or seeds (apomixis, cytoplasmic male sterility, plastid transformation, and Genetic Use Restriction Technology [GURT], etc.), though more can be done in this field.

While biotech research and development in Europe slowed significantly following the European Union's 1998 de facto moratorium on approvals and individual countries such as France and Germany bowing to internal minority pressure, Europe's stance on biotech crops cannot prevent biotech adoption in the rest of the world and, as noted, especially Asia is forging ahead. According to a study by Runge et al. (2005) as the EU becomes increasingly isolated, it will discourage its young scientists and technicians from pursuing European careers. They opine that if, on the other hand, the EU engages biotech in an orderly regulatory framework harmonized with the rest of the world, and it will encourage a more rapid international diffusion of the technology. More nations will join the top tiers of commercial production, and emerging nations will continue to expand the sector.

Conclusions

11

As agriculture must adapt to rapidly changing needs and growing conditions we must become more effective at producing more with limited resources, and only the tools of biotechnology will allow us to bypass physiological and environmental limitations to produce sufficient food, feed, fuels and fiber on ever diminishing arable land to meet ever increasing demand (sustainable intensification). The challenges going forward are foremost technical as we strive to modify qualitative as opposed to quantitative traits and intricate metabolic pathways and networks as opposed to single genes, the scientific hurdles to achieve these aims are not trivial. However, with the tools now coming on line in the fields of genomics, proteomics, metabolmics and bioinformatics, we have the potential to make major modifications to introgress desirable traits. For example, tools such as next generation sequencing, RNA interference (RNAi), transcription factors (Tfs), transcription activator-like effector nucleases (TALENs), mini-chromosomes, combinatorial transformation, epigenetic modification, network engineering and systems biology will allow us to apply both reductive and holistic approaches to identify, modify, introgress and subsequently simultaneously study the expression and interaction of transgenes on tens of thousands of endogenous genes in elite germplasm backgrounds. With these newly evolving tools, we are beginning to dissect the global effects of metabolic engineering on metabolites, enzyme activities and fluxes. With rapidly emerging technologies, the increase in our understanding of and ability to manipulate plant metabolism during the coming decades should place plant researchers in the position of being able to modify crop traits to respond to the diversity of needs from minimizing environmental impact to optimizing productivity and quality output.

Non-technical limitations include intellectual property restrictions, which may limit translation of public research if not managed judiciously; secondly, liability concerns over abuse or misuse of constructs; thirdly, prohibitive and asymmetric biosafety regimes and finally public acceptance. The latter two in many ways are the most insidious of limitations as they have little basis in rational process and thus are difficult to redress effectively – the last in particular is often predicated on how much of the former is perceived to be of concern, and how positions are presented by the opposing factions. It is often easier to appeal to emotion and sell fear than it is to present reasoned and judicious scientific rational for basing risk analysis. Indeed, the actual commercialization of biotech products may have little to do with technical limitations and more to do with these external constraints, primarily the process of regulatory approval. The flagship of improved nutritional varieties, namely beta carotene enhanced rice commonly referred to as golden rice, despite being under consideration since the late nineties and subject to a barrage of risk assessments did not receive approval until February 2013 in the Philippines. Ingo Potrykus, the developer, says an unreasonable amount of testing has been required without scientific justification. In a recent Nature article he lays the blame largely on the regulatory process, which he considers excessive observing that unjustified and impractical legal requirements are preventing genetically engineered crops from saving millions from starvation and malnutrition.

In the final analysis resources are finite and true sustainability can come only from an enlightened philosophy that promotes the development of resource-enhancing technologies. Antithetically, those who claim to be the stalwarts of sustainability are, on occasion, the very ones who oppose the development and application of those tools that can help to insure such sustainability. The only sure way to insure food security and protect the planet's resources is not to settle into the complacency of maintaining the status quo but to engage in continual, constructive change based on scientific knowledge. Thus, if we are to be accountable to posterity it is not just our choice but our duty to promote and apply responsible science and technology in all endeavors.

Glossary

Abiotic Stress: Outside (non-living) factors that can cause harmful effects to plants, such as soil conditions, drought, flooding, extreme temperatures.

Acclimatization: Adaptation of an organism to a new environment.

Adaptation: In the evolutionary sense, some heritable feature of an individual's phenotype that improves its chances of survival and reproduction in the existing environment.

Additive genetic variance: Genetic variance associated with the average effects of substituting one allele for another.

Agronomic performance/trait: Pertains to practices of agricultural production and its costs and the management of cropland. Examples of agronomic traits include yield, input requirements, stress tolerance.

Aldolase: An enzyme, not subject to allosteric regulation, that catalyzes in a reversible reaction the cleavage of fructose 1,6-biphosphate to form dihydroxyacetone phosphate and glyceraldehyde 3-phosphate. The enzyme catalyzing the fourth reaction in the glycolytic pathway, which splits a monosaccharide into two three-carbon units.

Agrobacterium tumefaciens: A bacterium normally responsible for production of crown gall disease in a variety of plants. A plasmid has been isolated from this bacterium that is useful in plant genetic engineering. This plasmid, called the *Ti* plasmid, has been modified so that it does not cause disease but can carry foreign DNA into susceptible plant cells.

Allelle: Any of several alternative forms of a given gene.

Allele frequency: Often called gene frequency. A measure of how common an allele is in a population; the proportion of all alleles at one gene locus that are of one specific type in a population.

Allelic exclusion: A process whereby only one immunoglobulin light chain and one heavy chain gene are transcribed in any one cell; the other genes are repressed.

Allogenic: Of the same species but with a different genotype.

Allopolyploid: Polyploid produced by the hybridization of two species.

Allopolyploid plants: Plants having more than two sets of haploid chromosomes inherited from different species.

Allosteric regulation: Regulation of an enzyme's activity by binding of a small molecule at a site that does not overlap the active site region.

Anabolic: That part of metabolism that is concerned with synthetic reactions.

Aneuploid: Having a chromosome number that is not an exact multiple of the haploid number, caused by one chromosome set being incomplete or chromosomes being present in extra numbers.

Aneuploidy: The condition of a cell or of an organism that has additions or deletions of a small number of whole chromosomes from the expected balanced diploid number of chromosomes.

Antibiotic: Chemical substance formed as a metabolic byproduct in bacteria or fungi and used to treat bacterial infections. Antibiotics can be produced naturally, using microorganisms, or synthetically.

Antibody: A protein produced by the immune system in response to an antigen (a molecule that is perceived to be foreign). Antibodies bind specifically to their target antigen to help the immune system destroy the foreign entity.

Antinutrients: Substances that act in direct competition with or otherwise inhibit or interfere with the use or absorption of a nutrient.

Antisense RNA: RNA produced by copying and reversing a portion of an RNA-encoding DNA, usually including a protein-specifying region, and placing it next to a transcription-control sequence. This cassette can be delivered to the target cell, resulting in genetic transformation and production of RNA that is complementary to the RNA that is produced from the original, not-reversed, DNA segment. This complementary, or antisense, RNA is able to bind to the complementary sequences of the target RNA, resulting in inhibition of expression of the target gene.

Antiserum: Blood serum containing specific antibodies against an antigen. Antisera are used to confer passive immunity to diseases and as analytical and preparative reagents for antigens for example to determine potential allergenicity.

Avirulent: Unable to cause disease.

Bacillus thuringiensis (Bt): A naturally occurring microorganism that produces a toxin protein that only kills organisms with alkalineing stomachs, such as insect larvae. This toxin protein has been used for biological control for decades. The genetic information that encodes the toxin protein was identified and moved into plants to make them insect-resistant.

Bioconversion: Chemical restructuring of raw materials by using a biocatalyst.

Biodegradable: Capable of being broken down by the action of microorganisms, usually by microorganisms and under conditions generally in the environment.

Bioinformatics: The discipline encompassing the development and utilization of computational facilities to store, analyze, and interpret biological data.

Biomass: The totality of biological matter in a given area. As commonly used in biotechnology, refers to the use of cellulose, a renewable resource, for the production of chemicals that can be used generate energy or as alternative feedstocks for the chemical industry to reduce dependence on nonrenewable fossil fuels.

Bioprocess: A process in which living cells, or components thereof, are used to produce a desired end product.

Biosynthesis: Production of a chemical by a living organism.

Biotechnology: Development of products by a biological process. Production may be carried out by using intact organisms, such as yeasts and bacteria, or by using natural substances (e.g., enzymes) from organisms.

Biosynthetic: relating to the formation of complex compounds formed from simple substances by living organisms.

Biotic stress: Living organisms that can harm plants, such as viruses, fungi, and bacteria, and harmful insects. See *Abiotic stress*.

Callus: A cluster of undifferentiated plant cells that can, for some species, be induced to form the whole plant.

Calvin cycle: A series of enzymatic reactions, occurring during photosynthesis, in which glucose is synthesized from carbon dioxide.

Catalyst: An agent (such as an enzyme or a metallic complex) that facilitates a reaction but is not itself changed at completion of the reaction.

Catabolic: That part of metabolism that is concerned with degradation reactions.

Chloroplast: A chlorophyll-containing photosynthetic organelle, found in eukaryotic cells, that can harness light energy.

Cistron: A length of chromosomal DNA representing the smallest functional unit of heredity, essentially identical to a gene.

Clone: A group of genes, cells, or organisms derived from a common ancestor. Because there is no combining of genetic material (as in sexual reproduction), the members of the clone are genetically identical or nearly identical to the parent.

Codon: A sequence of three nucleotide bases that in the process of protein synthesis specifies an amino acid or provides a signal to stop or start protein synthesis (translation).

Coenzyme: An organic compound that is necessary for the functioning of an enzyme. Coenzymes are smaller than the enzymes themselves and may be tightly or loosely attached to the enzyme protein molecule.

Cofactor: A non-protein substance required for certain enzymes to function. Cofactors can be coenzymes or metallic ions.

Comparative genomics: The comparison of genome structure and function across different species in order to further understanding of biological mechanisms and evolutionary processes.

Composition analysis: The determination of the concentration of compounds in a plant. Compounds that are commonly quantified are proteins, fats, carbohydrates, minerals, vitamins, amino acids, fatty acids and antinutrients.

Conventional breeding: Breeding of plants carried out by controlled transfer of pollen from one plant to another followed by selection of progeny through multiple generations for a desirable phenotype. This method has also often included irradiation or mutation of plants or seeds to induce extra variation in the donor material.

Complementary DNA (cDNA): DNA synthesized from an expressed messenger RNA through a process known as reverse transcription. This type of DNA is used for cloning or as a DNA probe for locating specific genes in DNA hybridization studies.

Crossbreeding: Interbreeding to breed (animals or plants) using parents of different races, varieties, breeds, etc.

Cyto: A prefix referring to cell or cell plasm.

Cytokines: Intercellular signals, usually protein or glycoprotein, involved in the regulation of cellular proliferation and function.

Diet: A specific allowance or selection of food or feed that a person or animal regularly consumes.

Diploid: A cell with two complete sets of chromosomes. *Cf* Haploid.

DNA sequencing: Technologies through which the order of base pairs in a DNA molecule can be determined.

Enzyme: A protein catalyst that facilitates specific chemical or metabolic reactions necessary for cell growth and reproduction. *Cf* Catalyst.

Epigenetics: The study of changes in gene expression caused by mechanisms other than changes in the underlying DNA sequence – hence the name epi- (Greek: επί: over, above, outer) genetics. Examples of such changes might be DNA methylation or histone deacetylation, both of which serve to suppress gene expression without altering the sequence of the silenced genes.

Event: The term used to describe a plant and its offspring that contain a specific insertion of DNA. Such events will be distinguishable from other events by their unique site of integration of the introduced DNA.

Exposure assessment: The qualitative and/or quantitative evaluation of the likely exposure to biological, chemical and physical agents via different sources.

Feedstock: The raw material used in chemical or biological processes.

Flavonoids: Any of a group of organic compounds that occur as pigments in fruit and flowers.

Food additive: Any substance not normally consumed as a food by itself and not normally used as a typical ingredient of food, whether or not it has nutritive value, the intentional addition of which to a food for a technological (including organoleptic) purpose in the manufacture, processing, preparation, treatment, packing, packaging, transport or holding of such food results, or may be expected to result (directly or indirectly), in it or its byproducts becoming a component of or otherwise affecting the characteristics of such foods. The term does not include "contaminants" or substances added to food for maintaining or improving nutritional qualities.

Fructan: A type of polymer of fructose, present in certain fruits.

Functional foods: The Institute of Medicine's Food and Nutrition Board defined functional foods as "any food or food ingredient that may provide a health benefit beyond the traditional nutrients it contains."

Functional genomics: The development and implementation of technologies to characterize the mechanisms through which genes and their products function and interact with each other and with the environment. This is usually applied to studies of gene the expression (mRNA) of usually large numbers of genes simultaneously.

Gene expression: The process through which a gene is activated at particular time and place so that its functional product is produced.

Gene flow: The exchange of genetic traits between populations by movement of individuals, gametes, or spores. It involves the spread of new variants among different populations through dispersal.

Gene silencing: (See RNAi.) A method usually performed by the expression of an mRNA of complementary or the same nucleotide sequence in a cell such that the expression of the mRNA causes the down regulation of the protein that is being targeted.

Gene transfer: The transfer of genes to an organism. Usually used in terms of transfer of a gene to an organism other that the original organism, through the tools of biotechnology.

Gene: A segment of chromosome that encodes the necessary regulatory and sequence information to direct the synthesis of a protein or RNA product.

Gene mapping: Determination of the relative locations of genes on a chromosome.

Gene sequencing: Determination of the sequence of nucleotide bases in a strand of DNA.

Genetic engineering: A technology used to alter the genetic material of living cells in order to make them capable of producing new substances or performing new functions.

Genetic map: A map showing the positions of genetic markers along the length of a chromosome relative to each other (genetic map) or in absolute distances from each other (physical map).

Genome: The total hereditary material of a cell, comprising the entire chromosomal set found in each nucleus of a given species.

Genome editing, or genome editing with engineered nucleases (GEEN): A type of genetic engineering in which bases or sequences are inserted, replaced, or removed from a genome using artificially engineered nucleases.

Genomics: Science that studies the genomes (i.e., the complete genetic information) of living beings. This commonly entails the analysis of DNA sequence data and the identification of genes.

Genotype: Genetic make-up of an individual or group. *Cf* Phenotype.

Germplasm: The total genetic variability, represented by germ cells or seeds, available within a particular population of organisms.

Gene pool: The total genetic information contained within a given population.

Glycoalkaloid toxins: Steroid-like compounds produced by plant members of the botanical family Solanaceae, most notably "solanine" present in potato tubers.

Golden Rice: In 1999, Swiss and German scientists announced the development of a genetically engineered rice crop that produces beta-carotene, a substance that the body converts to vitamin A. This improved nutrient rice was developed to treat individuals suffering from vitamin A deficiency, a condition that afflicts millions of people in developing countries, especially children and pregnant women.

Haploid: A cell with half the usual number of chromosomes, or only one chromosome set. Sex cells are haploid. *Cf* Diploid.

Hazard characterization: The qualitative and/or quantitative evaluation of the nature of the adverse health effects associated with biological, chemical, and physical agents. For chemical agents, a dose-response assessment should be performed if the data are obtainable.

Hazard identification: The identification of biological, chemical, and physical agents capable of causing adverse health or environmental effects.

Hazard: A biological, chemical, or physical agent, or condition, with the potential to cause an adverse health or environmental effect.

Hereditary: Capable of being transferred as genetic information from parent cells to progeny.

Heterozygote: With respect to a particular gene at a defined chromosomal locus, a heterozygote has a different allelic form of the gene on each of the two homologous chromosomes.

Homologous: Corresponding or alike in structure, position, or origin.

Homologous recombination: Rearrangement of related DNA sequences on a different molecule by crossing over in a region of identical sequence.

Homozygote: With respect to a particular gene at a defined chromosomal locus, a homozygote has the same allelic form of the gene on each of the two homologous chromosomes

Hormone: A chemical that acts as a messenger or stimulatory signal, relaying instructions to stop or start certain physiological activities. Hormones are synthesized in one type of cell and then released to direct the function of other cell types.

Horizontal gene transfer: Transmission of DNA between species, involving close contact between the donor's DNA and the recipient, uptake of DNA by the recipient, and stable incorporation of the DNA into the recipient's genome.

Host: A cell or organism used for growth of a virus, plasmid, or other form of foreign DNA, or for the production of cloned substances.

Hybridization: Production of offspring, or hybrids, from genetically dissimilar parents. The process can be used to produce hybrid plants (by cross-breeding two different varieties) or hybridomas (hybrid cells formed by fusing two unlike cells, used in producing monoclonal antibodies). The term is also used to refer to the binding of complementary strands of DNA or RNA.

Hybrid: The offspring of two parents differing in at least one genetic characteristic (trait). Also, a heteroduplex DNA or DNA-RNA molecule.

Identity preservation: The segregation of one crop type from another at every stage from production and processing to distribution. This process is usually performed through audits and site visits and provides independent third-party verification of the segregation.

Immunoassay: Technique for identifying substances based on the use of antibodies.

Immunogen: Any substance that can elicit an immune response, especially specific antibody production. An immunogen that reacts with the elicited antibody may be called an antigen.

Inbred: Progeny produced as a result of inbreeding.

Inducer: A molecule or substance that increases the rate of enzyme synthesis, usually by blocking the action of the corresponding repressor.

Inserted DNA: The segment of DNA that is introduced into the chromosome, plasmid or other vector using recombinant DNA techniques.

Introgressed: Backcrossing of hybrids of two plant populations to introduce new genes into a wild population.

Inulins:A fructose polysaccharide present in the tubers and rhizomes of some plants. Formula: (C6H10O5)n.

In vitro: Literally, "in glass." Performed in a test tube or other laboratory apparatus.

In vivo: In the living organism.

Invertase activity: Enzyme activity occurring in the intestinal juice of animals and in yeasts that hydrolyses sucrose to glucose and fructose.

Isoflavones: Water-soluble chemicals, also known as phytoestrogens, found in many plants and so named because they cause effects in the mammalian body somewhat similar to those of estrogen. The most investigated natural isoflavones, genistein and daidzen, are found in soy products and the herb red clover.

Knock in: Replacement of a gene by a mutant version of the same gene using homologous recombination.

Knock Out: Inactivation of a gene by homologous recombination following transfection with a suitable DNA construct.

Linkage: The tendency for certain genes to be inherited together due to their physical proximity on the chromosome.

Locus (plural loci): The position of a gene, DNA marker or genetic marker on a chromosome. See *gene locus*.

Macronutrient: Any substance, such as carbon, hydrogen, or oxygen, that is required in large amounts for healthy growth and development.

Marker: Any genetic element (locus, allele, DNA sequence or chromosome feature) that can be readily detected by phenotype, cytological or molecular techniques, and used to follow a chromosome or chromosomal segment during genetic analysis.

Marker assisted selection or marker aided selection (MAS): A process whereby a marker (morphological, biochemical, or based on DNA/RNA variation) is used for indirect selection of a genetic determinant or determinants of a trait of interest (i.e., productivity, disease resistance, abiotic stress tolerance, and/or quality). This process is used in plant and animal breeding.

Mass spectrometry: Analytical technique by which compounds in a vacuum compartment are ionized, eventually fragmented, accelerated, and detected based upon the mass-dependent behavior of the ionized compounds or their fragments in response to the application of a magnetic or electric field in a vacuum.

Messenger RNA (mRNA): Nucleic acid that carries instructions to a ribosome for the synthesis of a particular protein.

Metabolism: All biochemical activities carried out by an organism to maintain life.

Metabolite: A substance produced during or taking part in metabolism.

Metabolomics: "Open-ended" analytical techniques that generate profiles of the metabolites, i.e., chemical substances within a biological sample. Commonly, differences between profiles of different (groups of) samples are determined and the identity of the associated metabolites elucidated. Contrary to targeted analysis, these techniques are indiscriminate in that they do not require prior knowledge of every single substance that is present.

Microarray: A microscopic, ordered array of nucleic acids, proteins, small molecules, cells, or other substances that enables parallel analysis of complex biochemical samples. There are many different types of microarrays both from a biological and production system perspective. The generic terms "DNA array," "GeneChip," or "hybridization array" are used to refer broadly to all types of oligonucleotide-based arrays. The two most common are cDNA arrays and genomic arrays. cDNA array: A microarray composed of grid of nucleic acid molecules of known composition linked to a solid substrate, which can be probed with total messenger RNA from a cell or tissue to reveal changes in gene expression relative to a control sample.

Micronutrient: Any substance, such as a vitamin or trace element, essential for healthy growth and development but required only in minute amounts.

Mini-chromosome: Contains only centromeres and telomeres with little additional DNA. This provides the ability to accept multiple genes coding for stacked traits. They are particularly useful because they allow scientists to add numerous genes onto one mini-chromosome and manipulate those genes easily because they are all in one place.

mRNA: Messenger RNA.

Multigenic: Of hereditary characteristics, one that is specified by several genes.

Mutant: A cell that manifest new characteristics due to a change in its DNA.

Mutation: A structural change in a DNA sequence resulting from uncorrected errors during DNA replication.

Mutation Breeding: Genetic change caused by natural phenomena or by use of mutagens. Stable mutations in genes are passed on to offspring; unstable mutations are not.

Nitrogen fixation: A biological process (usually associated with plants) whereby certain bacteria convert nitrogen in the air to ammonia, thus forming a nutrient essential for growth.

Nucleic acid: Large molecules, generally found in the cell's nucleus and/or cytoplasm, that are made up of nucleotide bases. The two kinds of nucleic acid are DNA and RNA.

Nucleotides: The building blocks of nucleic acids. Each nucleotide is composed of sugar, phosphate, and one of four nitrogen bases. If the sugar is ribose, the nucleotide is termed a "ribonucleotide," whereas deoxyribonucleotides have deoxyribose as the sugar component (i.e., adenine, cytosine, guanine, and thymine in the case of DNA). The sequence of the nucleotides within the nucleic acid determines, for example, the amino acid sequence of an encoded protein.

Nucleus: In eukaryotic cells, the centrally-located organelle that encloses most of the chromosomes. Minor amounts of chromosomal substance DNA are found in some other organelles, most notably the mitochondria and the chloroplasts.

Nutritionally improved: Improving the quantity, ratio and/or bioavailability of essential macro and micronutrients and other compounds for which the clinical and epidemiological evidence is clear that they play a significant role in maintenance of optimal health and are limiting in diets.

Nutraceutical: The term was coined by the Foundation for Innovation in Medicine in 1991 and is defined as “any substance that may be considered a food or part of a food and provides medical or health benefits, including the prevention and treatment of disease.”

Organoleptic: Able to perceive a sensory stimulus such as taste.

Operon: Sequence of genes responsible for synthesizing the enzymes needed for biosynthesis of a molecule. An operon is controlled by an operator gene and a repressor gene.

Pathogen: Disease-causing organism.

Peptide: Two or more amino acids joined by a linkage called a peptide bond.

Pesticide: Any substance intended for preventing, destroying, attracting, repelling, or controlling any pest including unwanted species of plants or animals during the production, storage, transport, distribution, and processing of food, agricultural commodities, or animal feeds, or which may be administered to animals for the control of ectoparasites. The term includes substances intended for use as a plant-growth regulator, defoliant, desiccant, fruit-thinning agent, or sprouting inhibitor, and substances applied to crops either before or after harvest to protect the commodity from deterioration during storage and transport. The term normally excludes fertilizers, plant and animal nutrients food additives, and animal drugs.

Phenotype: Observable characteristics, resulting from interaction between an organism’s genetic make-up and the environment. *Cf* Genotype.

Phenylpropanoids: Especially the derivatives of the cinnamyl alcohols and of cinnamic acids, isolated from medicinal plants due to the interest as the source for the preparation of the remedies.

Photosynthesis: Conversion by plants of light energy into chemical energy, which is then used to support the plants’ biological processes.

Phytate (phytic acid): A phosphorus-containing compound in the outer husks of cereal grains that, in addition to limiting the bioavailability of phosphorous itself, binds with minerals and inhibits their absorption.

Phytochemicals: Small molecule chemicals unique to plants and plant products.

Plasmid: Circular extra-chromosomal DNA molecules present in bacteria and yeast. Plasmids replicate autonomously each time the organism a bacterium divides and are transmitted to the daughter cells. DNA segments are commonly cloned using plasmid vectors.

Plasticity: The quality of being plastic or able to be molded, changed.

Plastid: Any of various small particles in the cytoplasm of the cells of plants and some animals that contain pigments (see chromoplast), starch, oil, protein, etc.

Pleiotropic: Genes or mutations that result in the production of multiple effects at the phenotypic level. It is the consequence of the fact that biochemical pathways starting from different genes intersect in many places, inhibiting, deflecting, and variously modifying each other. Introduced genes may also insert into sites that effect phenotypic changes other than the one desired.

Polyclonal: Derived from different types of cells.

Polymer: A long molecule of repeated subunits.

Polypeptide: Long chain of amino acids joined by peptide bonds.

Post-Transcriptional Gene Silencing: is a sequence-specific RNA degradation system designed to act as an anti-viral defense mechanism. A form of PTGS triggered by transgenic DNA, called co-suppression, was initially described in plants, and a related phenomenon, termed quelling, was later observed in the filamentous fungus Neurosporacrassa.

Post-transcriptional modification: A process through which protein molecules are biochemically modified within a cell following their synthesis by translation of messenger RNA. A protein may undergo a complex series of modifications in different cellular compartments before its final functional form is produced.

Profiling: Creation of indiscriminate patterns of the substances within a sample with the aid of analytical techniques, such as functional genomics, proteomics, and metabolomics. The identity of the compounds detectable within the pattern need not be known.

Promoter: A DNA sequence that is located near or even partially within encoding nucleotide sequences and which controls gene expression. Promoters are required for binding of RNA polymerase to initiate transcription.

Protein: Proteins are biological effector molecules encoded by an organism's genome. A protein consists of one or more polypeptide chains of amino acid subunits. The functional action of a protein depends on its three-dimensional structure, which is determined by its amino acid composition and any post-transcriptional modifications.

Proteomics: The development and application of techniques used to investigate the protein products of the genome and how they interact to determine biological functions. This is an "open-ended" analytical technique that generates profiles of the proteins within a biological sample. Commonly, it is used to find differences between profiles of different (groups of) samples, and the identity of the associated proteins is elucidated. Contrary to targeted analysis, these techniques are indiscriminate in that they do not require prior knowledge of every single protein present.

Protoplast fusion: The fusion of two plant protoplasts that each consist of the living parts of a cell, including the protoplasm and cell membrane but not the vacuoles or the cell wall.

Protoplast: The cellular material that remains after the cell wall has been removed. A plant cell from which the cell wall has been removed by mechanical or enzymatic means. Protoplasts can be prepared from primary tissues of most plant organs as well as from cultured plant cells.

Quantitative trait loci: The locations of genes that together govern a multigenic trait, such as yield or fruit mass.

Recombinant DNA: Any DNA molecule formed by joining DNA segments from different sources (not necessarily different organisms). This may also be a strand of DNA synthesized in the laboratory by splicing together selected parts of DNA strands from different organic species, or by adding a selected part to an existing DNA strand.

Regeneration: Laboratory technique for forming a new plant from a clump of plant cells.

Regulatory gene: A gene that acts to control the protein-synthesizing activity of other genes.

Regulatory sequence: A DNA sequence to which specific proteins bind to activate or repress the expression of a gene.

Replication: Reproduction or duplication, as of an exact copy of a strand of DNA.

Rhizobium: A class of microorganisms that converts atmospheric nitrogen into a form that plants can utilize for growth. Species of this microorganism grow symbiotically on the roots of certain legumes such as peas, beans, and alfalfa.

Ribonucleic acid (RNA): A molecule similar to DNA that functions primarily to decode the instructions for protein synthesis that are carried by genes. See also *Messenger RNA; transfer RNA*.

Ribosome: A cellular component, containing protein and RNA, that is involved in protein sythesis.

Ribozyme: Any of the RNA molecules possessing catalytic activity and acting as biological catalysts.

Risk: A function of the probability of an adverse health effect and the severity of that effect, consequential to a hazard(s).

Risk analysis: A process consisting of three components: risk assessment, risk management, and risk communication.

Risk assessment: A scientific process consisting of the following steps: (i) hazard identification, (ii) hazard characterization, (iii) exposure assessment, and (iv) risk characterization.

Risk characterization: The qualitative and/or quantitative estimation, including attendant uncertainties, of the probability of occurrence and severity of known or potential adverse health effects in a given population based on hazard identification, hazard characterization, and exposure assessment.

Risk communication: The interactive exchange of information and opinions throughout the risk analysis process concerning hazards and risks, risk-related factors, and risk perceptions, among risk assessors, risk managers, population, industry, the academic community, and other parties, including the explanation of risk assessment findings and the basis of risk management decisions.

Risk management: The process, distinct from risk assessment, of weighing policy alternatives, in consultation with all interested parties, considering risk assessment and other factors relevant for the health protection of population and for the promotion of fair practices, and if needed, selecting appropriate prevention and control options.

RNAi: RNA Interference (RNAi), a term coined by Fire et al in 1998, is a phenomenon whereby small double-stranded RNA (referred as small interference RNA or siRNA) can induce efficient sequence-specific silence of gene expression.

SAFOTEST: EU project on new methods for the safety testing of transgenic food.

Scale-up: Transition from small-scale production to production of large industrial quantities.

Secondary metabolites: Chemical substances within a biological organism sample that are not necessary for concerned with primary cellular functions. Secondary metabolism proceeds by modification of the primary metabolites of photosynthesis, respiration, etc. by four main pathways. The malonate/polyketide pathway leads to the production of fatty acids and naphthoquinones. The mevalonate/isoprenoid pathway leads to the various terpenes (such as menthol), carotenoids and steroids. The shikimate pathway leads to aromatic amino acids and the phenolics and the final group of metabolites is a non-specific mix of amino-acid derivatives including the alkaloids (such as solanine) and others of mixed biogenesis.

Selectable marker: A gene, often encoding resistance to an antibiotic or an herbicide, introduced into a group of cells to allow identification of those cells that contain the gene of interest from the cells that do not. Selectable markers are used in genetic engineering to facilitate identification of cells that have incorporated another desirable trait that is not easy to identify in individual cells.

Selective breeding: Making deliberate crosses or matings of organisms so the offspring will have particular desired characteristics derived from one or both of the parents.

Selective medium: Nutrient material constituted such that it will support the growth of specific organisms while inhibiting the growth of others.

Sequence homology: The measurable degree of identity or similarity between two nucleotide or amino acid sequences.

Sera-binding tests: Immunological assays that evaluate for the presence of antigen-specific IgE in blood serum obtained from individuals allergic to food, pollen, or other environmental antigens. Sera-binding tests include assays such as western blotting, ELISA, ELISA-inhibition, RAST and RAST-inhibition techniques.

Shikimate pathway: Pathway in micro-organisms and plants involved in the biosynthesis of the aromatic amino acid family (phenylalanine, tyrosine, tryptophan) with a requirement for chorismate as well as shikimate. Secondary metabolites such as lignin, pigments, UV light protectants, phenolic redox molecules, and other aromatic compounds such as folic acid and ubiquinone are postscript products of the shikimate pathway.

Signal transduction: The molecular pathways mechanism through which a cell senses changes in its external environment and changes its gene expression patterns in response.

Signal sequence: The N-terminal sequence of a secreted protein, which is required for transport through the cell membrane.

Small Interfering RNA (siRNA): Small Interfering RNA (siRNA) is 21~23-nt double-stranded RNA molecules. It guides the cleavage and degradation of its cognate RNA.

Site-specific recombination: A crossover event, such as the integration of phage lambda, that requires homology of only a very short region and uses an enzyme specific for that recombination. Recombination occurring between two specific sequences that need not be homologous; mediated by a specific recombination system.

Somaclonal selection: Epigenetic or genetic changes, sometimes expressed as a new trait, resulting from in vitro culture of higher plant cells. Somatic (vegetative non-sexual) plant cells can be propagated in vitro in an appropriate nutrient medium. The cells that multiply by division of the parent somatic cells are called somaclones and, theoretically, should be genetically identical with the parent. In fact this process frequently generates cells plants that are significantly different, epigenetically and/or genetically, from the parent in a stable fashion. Such progeny are called somaclonal variants and may provide a useful source of genetic variation.

Stilbenes: A colorless or slightly yellow crystalline water-insoluble unsaturated hydrocarbon used in the manufacture of dyes; trans-1,2-diphenylethene. Formula: C6H5CH:CHC6H5. It forms the backbone structure of several compounds with estrogenic activity. Trans-3,4',5-trihydroxy-stilbene, also known as resveratrol, has been found in some experiments to inhibit cell mutations, stimulate at least one enzyme that can inactivate certain carcinogens, and may contribute to a low incidence of cardiovascular disease.

Structural gene: A gene that codes for a protein, such as an enzyme.

Substantial equivalence: In the report of the 1996 FAO/WHO Expert Consultation, substantial equivalence was identified as being "established by a demonstration that the characteristics assessed for the genetically modified organism, or the specific food product derived therefrom, are equivalent to the same characteristics of the conventional comparator. The levels and variation for characteristics in the genetically modified organism must be within the natural range of variation for those characteristics considered in the comparator and be based upon an appropriate analysis of data." In the Codex Guideline for the Conduct of Food Safety Assessment of Foods Derived from Recombinant-DNA Plants (2003), the concept of substantial equivalence is described as "a key step in the safety assessment process. However, it is not a safety assessment in itself; rather it represents the starting point that is used to structure the safety assessment of a new food relative to its conventional counterpart. This concept is used to identify similarities and differences between the new food and its conventional counterpart. It aids in the identification of potential safety and nutritional issues and is considered the most appropriate strategy to date for safety assessment of foods derived from recombinant-DNA plants. The safety assessment carried out in this way does not imply absolute safety of the new product; rather, it focuses on assessing the safety of any identified differences so that the safety of the new product can be considered relative to its conventional counterpart."

Substrate: Material acted on by an enzyme.

Synteny: All loci on one chromosome are said to be syntenic (literally on the same ribbon). Loci may appear to be unlinked by conventional genetic tests for linkage but still be syntenic.

Systems biology: A biology-based inter-disciplinary study field that focuses on complex interactions in biological systems, purported to be a new perspective (holism instead of reduction). Particularly from the year 2000 onwards, the term is used widely in the biosciences, and in a variety of contexts. An oft-stated ambition of systems biology is the modeling and discovery of emergent properties, properties of a system whose theoretical description is only possible using techniques that fall under the remit of systems biology.

Tannins: Any of a class of yellowish or brownish solid compounds found in many plants and used as tanning agents, mordants, medical astringents, etc. Tannins are derivatives of gallic acid with the approximate formula C76H52O46.

T-DNA: The segment of the Ti plasmid of A. tumefaciens that is transferred to the plant genome following infection.

Ti plasmid: A plasmid containing the gene(s) responsible for inducing plant tumor formation.

Tissue culture: In vitro growth in nutrient medium of cells isolated from tissue.

Traditional breeding: Modification of plants and animals through selective breeding. Practices used in traditional plant breeding may include aspects of biotechnology such as tissue culture and mutational breeding.

Transcription: The process through which a gene is expressed to generate a complementary messenger RNA molecule. Synthesis of messenger (or any other) RNA on a DNA template.

Transcription Activator-Like Effector Nucleases (TALENs): Transcription activator–like effector (TALE) proteins from Xanthomonasare nucleases that cleave unique genomic sequences in living cells and can be used for targeted gene editing and mutagenesis.

Transcriptome: The total messenger RNA expressed in a cell or tissue at a given point in time.

Transgene: A gene from one source that has been incorporated into the genome of another organism.

Transgenic plant: A fertile plant that carries an introduced gene(s) in its germ-line.

Transformation: Change in the genetic structure of an organism by the incorporation of foreign DNA.

Transgenic organism: An organism formed by the insertion of foreign genetic material into the germ line cells of organisms. Recombinant DNA techniques are commonly used to produce transgenic organisms.

Translation: Process by which the information on a messenger RNA molecule is used to direct the synthesis of a protein.

Transmissible spongiform encephalopathy: A disease that can be transmitted from one animal to another and will produce changes in the brain that appear similar to a sponge (i.e., some of the cells are clear when seen down the microscope).

Transposon: A segment of DNA that can move around and be inserted at several sites in the genome of a cell possibly altering expression. The first to be described was the Ac/Ds system in maize shown by McClintock to cause unstable mutations.

Trypsin inhibitors: Antinutrient proteins present in plants such as soybeans that inhibit the digestive enzyme trypsin if not inactivated by heating or other processing methods.

Unintended effect: An effect that was not the purpose of the genetic modification or mutation. An unintended effect may be either predictable or unpredictable, based on the knowledge of, among other things, the function of the introduced DNA and of the native DNA affected by the genetic modification. A predicted unintended effect would be for example variations in metabolic intermediates and endpoints; an unpredicted effect might be turning on of unknown endogenous genes.

Variety: A subdivision of a species for taxonomic classification also referred to as a "cultivar." A variety is a group of individual plants that is uniform, stable, and distinct genetically from other groups of individuals in the same species.

Virulence: Ability to infect or cause disease.

Virus: A submicroscopic organism that contains genetic information but cannot reproduce itself. To replicate, it must invade another cell and use parts of that cell's reproductive machinery.

Wildtype: The form of an organism that occurs most frequently in nature.

Sources

Abbadi, A.D.F. et al. (2004) Biosynthesis of very-long-chain polyunsaturated fatty acids in transgenic oilseeds: constraints on their accumulation. Plant Cell 16, 2734–2748

Agbios 2008 GM crop database. Agbios. Available from: http://www.agbios.com/dbase.php?action=ShowForm.

Agius, F. et al. (2003) Engineering increased vitamin C levels in plants by overexpression of a D-galacturonic acid reductase. Nat. Biotechnol. 21, 177–181

Amyris, 2013 Amyris Biofuel research http://www.amyris.com/Innovation/155/BreakthroughScience

Anai, T. et al. (2003) Improvement of rice (Oryza sativa L.) seed oil quality through introduction of a soybean microsomal omega-3 fatty acid desaturase gene. Plant Cell Rep. 21, 988–992

Arcadia Biosciences and Bioriginal Food and Science Corp. 2008 Enter Strategic Alliance to Market High GLA Safflower Oil. Business Wire, Feb 22, 2008. Available at: http://www.arcadiabio.com/news/press-release/arcadia-biosciences-and-bioriginal-food-and-science-corp-enter-strategic-alliance

Arcadia Biosciences, (2008) Arcadia Biosciences and Bioriginal Food and Science Corp. Enter Strategic Alliance to Market High GLA Safflower Oil. Business Wire. Available from: http://findarticles.com/p/articles/mi_m0EIN/is_2008_Feb_22/ai_n24320185/

Arcadia, NUE 2013 Nitrogen Use efficiency, http://www.arcadiabio.com/nitrogen

Atanassov, A.B.A. et al. (2004) To reach the poor – results from the ISNAR-IFPRI Next Harvest study on genetically modified crops, public research, and policy implications. EPTD Discussion Paper – Environment and Production Technology Division International Food Policy Research Institute

Austin-Phillips, S.B.E. et al. (1999) Production of Industrial and Animal Feed Enzymes in Transgenic Alfalfa. Available from: http://www.molecularfarming.com/nonmedical.html

Brinch-Pedersen, H. et al. (2000) Generation of transgenic wheat (Triticum aestivum L.) for constitutive accumulation of an Aspergillus phytase. Mol. Breed. 6, 195–206

Brookes, G., & Barfoot P. (2012a) The income and production effects of biotech crops globally 1996–2010. GM Crops and Food: Biotechnology in Agriculture and the Food Chain 3:265 - 272; PMID: 22750951; http://dx.doi.org10.4161/gmcr.20097

Bruening, G and Lyons, J.M. (2000) The case of the FLAVR SAVR tomato. California Agriculture 54(4):6-7. DOI: 10.3733/ca.v054n04p6. July-August 2000

Cahoon, E.B. et al. (2003) Metabolic redesign of vitamin E biosynthesis in plants for tocotrienol production and increased antioxidant content. Nat. Biotechnol. 21, 1082–1087

Caimi, P.G. et al. (1996) Fructan accumulation and sucrose metabolism in transgenic maize endosperm expressing a Bacillus amyloliquefaciens SacB gene. Plant Physiol. 110, 355–363

Chakraborty, S. et al. (2000) Increased nutritive value of transgenic potato by expressing a nonallergenic seed albumin gene from Amaranthus hypochondriacus. Proc. Natl. Acad. Sci. U. S. A. 97, 3724–3729

Chapman, K.D.A.-B.S. et al. (2001) Transgenic cotton plants with increased seed oleic acid content. J. Am. Oil Chem. Soc. 78, 941–947

Chassy, B. et al. (2004) Nutritional and safety assessments of foods and feeds nutritionally improved through biotechnology – Prepared by a task force of the ILSI International Food Biotechnology Committee. Comprehensive Reviews in Food Science and Food Safety 3, 35-104

Chassy, B. et al. (2008) Recent developments in the safety and nutritional assessment of nutritionally improved foods and feeds. Comp. Rev. Food Sci. Food Safety 7, 50-113

Chen, Z. et al. (2003) Increasing vitamin C content of plants through enhanced ascorbate recycling. Proc. Natl. Acad. Sci. U. S. A. 100, 3525–3530

Chiang, C.Y.F. et al. (2005) Expression of a bi-functional and thermostable amylopullulanase in transgenic rice seeds leads to autohydrolysis and altered composition of starch. Mol. Breed. 15, 125–143

Choudhary, B. & Gaur, K. 2008. The Development and Regulation of BT Brinjal in India (Eggplant/Aubergine). ISAAA38.

Choudhary, B. & Gaur, K. 2010. BT Cotton in India: A Country Profile ISAAA.

Codex Alimentarius (2003) Codex Principles and Guidelines on Foods derived from Biotechnology. http://ftp.fao.org/codex/standard/en/CodexTextsBiotechFoods.pdf

Cockburn, A. (2002) Assuring the safety of genetically modified (GM) foods: the importance of an holistic, integrative approach. Journal of Biotechnology 98, 79-106

Crispeels, M.J. & Sadava, D.E. (2003) Plants, Genes, and Crop Biotechnology, 2nd Ed., Jones and Bartlett Publishers

Cromwell, G.L. et al. (1967) Nutritional value of opaque-2 corn for swine. J. Anim. Sci. 26, 1325–1331

Del Vecchio, A. (1996) High-laurate canola. How Calgene's program began, where it's headed. INFORM – Int. News Fats Oils Relat. Mater. 7:, 230–243

Denbow, D.M. et al. (1998) Soybeans transformed with a fungal phytase gene improve phosphorus availability for broilers. Poult. Sci. 77, 878–881

Diaz de la Garza, R. et al. (2004) Folate biofortification in tomatoes by engineering the pteridine branch of folate synthesis. Proc. Natl. Acad. Sci. U. S. A. 101, 13720–13725

Dinkins, R.D.R.M. et al. (2001) Increased sulfur amino acids in soybean plants overexpressing the maize 15 kDa zein protein. In Vitro Cell. Dev. Biol. Plant 37, 742–747

Dodo, H.W., et al (2008) Alleviating peanut allergy using genetic engineering: the silencing of the immunodominant allergen Ara h 2 leads to its significant reduction and a decrease in peanut allergenicity. Plant Biotechnol J. 135-45

Drakakaki, G. et al. (2005) Endosperm-specific co-expression of recombinant soybean ferritin and Aspergillus phytase in maize results in significant increases in the levels of bioavailable iron. Plant Mol. Biol. 59, 869–880

Ducreux, L.J. et al. (2005) Metabolic engineering of high carotenoid potato tubers containing enhanced levels of beta-carotene and lutein. J. Exp. Bot. 56, 81–89

Dykxhoorn, D.M., Novina, C.D. and Sharp, P.A. 2003. Killing the Messenger: Short RNAs that Silence Gene Expression. Nature Revs. Mol. Cell. Biol. 4: 457-467

Enfissi, E.M. et al. (2005) Metabolic engineering of the mevalonate and nonmevalonate isopentenyl diphosphate-forming pathways for the production of health-promoting isoprenoids in tomato. Plant Biotechnol. J. 3, 17–27

Esvelt, KM.; Wang, HH. (2013). Genome-scale engineering for systems and synthetic biology. Mol Syst Biol 9: 641.

Europa Press Release (2010) Commission publishes compendium of results of EU-funded research on genetically modified crops http://europa.eu/rapid/press-release_IP-10-1688_en.htm Accessed March 17, 2013

Falco, S.C. et al. (1995) Transgenic canola and soybean seeds with increased lysine. Biotechnology (N. Y.) 13, 577–582

FAO (2013) at: http://www.fao.org/publications/sofi/en/

Fitch, M.M., Manshardt, R.M., Gonsalves, D., Slightom, J.L., & Sanford, J.C. 1992. Virus resistant papaya derived from tissues bombarded with the coat protein gene of papaya ringspot virus. Bio/Technology, 10, 1466-1472.

Fraser, P.D.R.S. et al. (2001) Elevation of carotenoids in tomato by genetic manipulation. J. Sci. Food Agric. 81, 822–827

Froman, B. (2002) Genetic modification of oils for improved health benefits: production of long chain omega-3 fatty acids in plants. Abstr. Papers Am. Chem. Soc. 223 U35-U

Galili, G. et al. (2002) Genetic, molecular, and genomic approaches to improve the value of plant foods and feeds. Crit. Rev. Plant Sci. 21, 167–204

Gan, S. & Amasino R.M. 1997 Making sense of senescence: molecular genetic regulation of leaf senescence. Plant Physiol 113:313–319.

Giovinazzo, G. et al. (2005) Antioxidant metabolite profiles in tomato fruit constitutively expressing the grapevine stilbene synthase gene. Plant Biotechnol. J. 3, 57–69

Glei, M. et al. (2006) Both wheat (Triticum aestivum) bran arabinoxylans and gut flora-mediated fermentation products protect human colon cells from genotoxic activities of 4-hydroxynonenal and hydrogen peroxide. J. Agric. Food Chem. 54, 2088–2095

Gonsalves, D., Tripathi S., Carr J.B., & Suzuki J.Y. (2010) Papaya Ringspot virus. The Plant Health Instructor. DOI: 10.1094/PHI-I-2010-1004-01

Gonzali, S. et al. (2009) Purple as a tomato: towards high anthocyanin tomatoes. Trends Plant Sci. 14, 237–241

Gosal, S.S., Wani S.H., Kang, M.S. (2009) Biotechnology and drought tolerance. J. Crop Improvement, 23: 19-54

Goto, F. et al. (2000) Iron accumulation and enhanced growth in transgenic lettuce plants expressing the iron-binding protein ferritin. Theor. Appl. Genet. 100, 658–664

Han, G. (2009) Origin Agritech Announces Final Approval of World's First Genetically Modified Phytase Corn. Available from: http://www.allbusiness.com/science-technology/biology-biotechnology-genetic/13453842-1.html

Hartwig, E. et al. (1997) Seed protein and its relationship to soluble sugars in soybeans. Crop Sci. 37, 770–773

Hellwege, E.M. et al. (1997) Transgenic potato tubers accumulate high levels of 1-kestose and nystose: functional identification of a sucrose 1- fructosyltransferase of artichoke (Cynara scolymus) blossom discs. Plant J. 12, 1057–1065

Hellwege, E.M. et al. (2000) Transgenic potato (Solanum tuberosum) tubers synthesize the full spectrum of inulin molecules naturally occurring in globe artichoke (Cynara scolymus) roots. Proc. Natl. Acad. Sci. U. S. A. 97, 8699–8704

Hipskind, J.D. & Paiva, N.L. (2000) Constitutive accumulation of a resveratrol-glucoside in transgenic alfalfa increases resistance to Phoma medicaginis. Mol. Plant Microbe Interact. 13, 551–562

Jalani, B.S.C.S. et al. (1997) Improvement of palm oil through breeding and biotechnology. J. Am. Oil Chem. Soc. 74, 1451–1455

James C. (2013) Global Status of Commercialized Biotech/GM Crops: 2012 ISAAA, Brief No. 44. Ithaca, N.Y.: International Service for the Acquisition of Agri-biotech Applications. Available from: http://www.isaaa.org/. Accessed 2012 Jan 15

James, M.J. et al. (2003) Metabolism of stearidonic acid in human subjects: comparison with the metabolism of other n-3 fatty acids. Am. J. Clin. Nutr. 77, 1140–1145

Kalaitzandonakes, N. et al. (2007) Compliance costs for regulatory approval of new biotech crops. Nature Biotech. 25, 509-511

Kalaitzandonakes, N. & Bijman, J. (2003) Who is driving biotechnology acceptance? Nature Biotechnology 21, 366-369

Katsube, T. et al. (1999) Accumulation of soybean glycinin and its assembly with the glutelins in rice. Plant Physiol. 120, 1063–1074

Kinney, A.J. (1998) Designer Oils: The High Oleic Acid Soybean (Roller, S.H.S., ed.), Blackie Academic and Professional

Knothe, Gerhard (2007) The Biodiesel Handbook, Chaper 2 – The History of Vegetable Oil Based Diesel Fuels, by Gerhard Knothe, ISBN 978-1-893997-79-0

Kobayashi, S.D.C. et al. (2000) Kiwifruits (Actinidia deliciosa) transformed with a Vitis stilbene synthase gene produce piceid (resveratrol-glucoside). Plant Cell Rep. 19, 904–910

Kok, E.J. & Kuiper, H.A. (2003) Comparative safety assessment for biotech crops. Trends in Biotechnology 21, 439-444

Konig, A. et al. (2004) Assessment of the safety of foods derived from genetically modified (GM) crops. Food and Chemical Toxicology 42, 1047-1088

Lai, J.S. (2002) Increasing maize seed methionine by mRNA stability. Plant J. 30, 395–402

Li, L.L.S. et al. (2001) Increase of sulphur-containing amino acids in transgenic potato with 10 ku zein gene from maize. Chin. Sci. Bull. 46, 482–484

Liu, Q. et al. (2002) High-oleic and high-stearic cottonseed oils: nutritionally improved cooking oils developed using gene silencing. J. Am. Coll. Nutr. 21 (3 Suppl.), 205S–211S

Lucca, P. et al. (2002) Fighting iron deficiency anemia with iron-rich rice. J. Am. Coll. Nutr. 21 (3 Suppl.), 184S–190S

Luciani, G. & Wofford, D.S. (2005) Over-Expression of a Soybean Vegetative Storage Protein Gene in Bahiagrass (Paspalum notatum var. Flugge). Available from: http://www.ufgi.ufl.edu/Symposium/FG2005program_rev.pdf

Lukaszewicz, M. et al. (2004) Antioxidant capacity manipulation in transgenic potato tuber by changes in phenolic compounds content. J. Agric. Food Chem. 52, 1526–1533

Mathers, J.C. (2006) Plant foods for human health: research challenges. Proc. Nutr. Soc. 65, 198-203

McCabe, M.S., Garratt, L.C., Schepers, F., Jordi, W.J.R.M., Stoopen, G.M., Davelaar, E., van Rhijn, H.A., Power, J.B., & Davey, M.R. (2001) Effects of PSAG12-IPT Gene Expression on Development and Senescence in Transgenic Lettuce Plant Physiology October 2001 vol. 127 no. 2, 505-516

McDougall, P. (2011) The cost and time involved in the discovery, development and authorization of a new plant biotechnology derived trait, A Consultancy Study for Crop Life International

Muir, S.R. et al. (2001) Overexpression of petunia chalcone isomerase in tomato results in fruit containing increased levels of flavonols. Nat. Biotechnol. 19, 470–474

Naqvi, S. et al. (2009) Transgenic multivitamin corn through biofortification of endosperm with three vitamins representing three distinct metabolic pathways. Proc. Nat. Acad. Sci. U. S. A. 106, 7762–7767

Niggeweg, R. et al. (2004) Engineering plants with increased levels of the antioxidant chlorogenic acid. Nat. Biotechnol. 22, 746–754

Novozymes Biofuels 2013 http://www.bioenergy.novozymes.com/en/the-basics/Pages/default.aspx

O'Quinn, P.R. et al. (2000) Nutritional value of a genetically improved highlysine, high-oil corn for young pigs. J. Anim. Sci. 78, 2144–2149

Parrot, W. (2005) The Nature of Change: Towards Sensible Regulation of Transgenic Crops Based on Lessons from Plant Breeding, Biotechnology and Genomics. National Agricultural Biotechnology Council Report 17, Agricultural Biotechnology: Beyond Food and Energy to Health and the Environment, 209-220

Parrott W., et al (2010) Application of food and feed safety assessment principles to evaluate transgenic approaches to gene modulation in crops. Food Chem Toxicol 48: 1773–1790

Parveez, G. (2003) Novel products from transgenic oil palm. AgBiotechNet 5, 1–8

Prakash, C.S. & Jaynes, J. (2000) Increasing the protein content in sweet potato using a synthetic storage protein gene. Abstr. Papers Am. Chem. Soc. 219, U36

Rapp, W. (2002) Development of soybeans with improved amino acid composition. 93rd AOCS Annual Meeting & Expo pp. 79–86, American Oil Chemists' Society Press

Reddy, A.S. & Thomas, T.L. (1996) Expression of a cyanobacterial delta 6- desaturase gene results in gamma-linolenic acid production in transgenic plants. Nat. Biotechnol. 14, 639–642

Redenbaugh, K. & McHughen, A. (2004) Regulatory Challenges Reduce Opportunities for Horticultural Crops. California Agriculture 58, 106-115

Ricroch, A.E. et al. (2011) Evaluation of Genetically Engineered Crops Using Transcriptomic, Proteomic, and Metabolomic Profiling Techniques. Plany Physiology 155, 1752-1761

Rocheford, T.R. et al. (2002) Enhancement of vitamin E levels in corn. J. Am. Coll. Nutr. 21 (3 Suppl.), 191S–198S

Roesler, K. et al. (1997) Targeting of the Arabidopsis homomeric acetyl-coenzyme A carboxylase to plastids of rapeseeds. Plant Physiol. 113, 75–81

Rosati, C. et al. (2000) Metabolic engineering of beta-carotene and lycopene content in tomato fruit. Plant J. 24, 413–419

Schwall, G.P. et al. (2000) Production of very-high-amylose potato starch by inhibition of SBE A and B. Nat. Biotechnol. 18, 551–554

Scott, S.A., Davey, M.P., Dennis, J.S., Horst, I., Howe, C.J., Lea-Smith, D.J., Smith, A.G. (2010) Biodiesel from algae: challenges and prospects. Current Opinion in Biotechnology. 21(3):277-86

Sevenier, R. et al. (1998) High level fructan accumulation in a transgenic sugar beet. Nat. Biotechnol. 16, 843–846

Shewmaker, C.K. et al. (1999) Seed-specific overexpression of phytoene synthase: increase in carotenoids and other metabolic effects. Plant J. 20, 401– 412

Shintani, D. & DellaPenna, D. (1998) Elevating the vitamin E content of plants through metabolic engineering. Science 282, 2098–2100

Smeekens, S. (1997) Engineering plant metabolism. Trends Plant Sci. 2, 286–287

Society of Toxicology (2003) The Safety of Genetically Modified Foods Produced through Biotechnology. Toxicological Sciences 71, 2-8

Sottosanto, J.B., Saranga, Y., Blumwald, E. (2007) Impact of AtNHX1, a vacuolar Na+/H+ antiporter, upon gene expression during short and long term salt stress in Arabidopsis thaliana. BMC Plant Biology. 7:18.

Sprenger, N. et al. (1997) Fructan synthesis in transgenic tobacco and chicory plants expressing barley sucrose: fructan 6-fructosyltransferase. FEBS Lett. 400, 355–358

Stark-Lorenzen, P.N.B. et al. (1997) Transfer of a grapevine stilbene synthase gene to rice (Oryza sativa L.). Plant Cell Rep. 16, 668–673

Stearidonic Acid (SDA) 2011, Monsanto Omega-3 Soybeans, http://www.monsanto.com/products/Pages/sda-omega-3-soybeans.aspx)

Steiner, H-Y, et al. (2013), 'Editor's Choice: Evaluating the potential for adverse interactions within genetically engineered breeding stacks' Plant Physiology, vol 161, no. 4, pp. 1587-1594.

Szankowski, I. et al. (2003) Transformation of apple (Malus domestica Borkh.) with the stilbene synthase gene from grapevine (Vitis vinifera L.) and a PGIP gene from kiwi (Actinidia deliciosa). Plant Cell Rep. 22, 141–149

Uauy, C. et al. (2006) Gene regulating senescence improves grain protein, zinc, and iron content in wheat. Science 314, 1298–1301

UPI United Press International, (2002) Wheat Inhibits Colon Cancer.

Ursin, V.M. (2003) Modification of plant lipids for human health: development of functional land-based omega-3 fatty acids. J. Nutr. 133, 4271–4274

USDA APHIS, 2012 http://www.aphis.usda.gov/biotechnology/petitions_table_pending.shtml#not_reg

Vistive Gold Soybeans, http://www.monsanto.com/products/Pages/vistive-gold-soybeans.aspx

Weber N, Halpin C, Hannah LC, Jez JM, Kough J, Parrott W, (2012) Editor's choice: crop genome plasticity and its relevance to food and feed safety of genetically engineered breeding stacks. Plant Physiol 160: 1842–1853

White, C.L.T.L. et al. (2001) Increased efficiency of wool growth and live weight gain in Merino sheep fed transgenic lupin seed containing sunflower albumin. J. Sci. Food Agric. 81, 147–154

Yang, S.H. et al. (2002) Expression of a synthetic porcine alpha-lactalbumin gene in the kernels of transgenic maize. Transgenic Res. 11, 11–20

Ye, X. et al. (2000) Engineering the provitamin A (beta-carotene) biosynthetic pathway into (carotenoid-free) rice endosperm. Science 287, 303– 305

Young, T.E. et al. (2004) Senescence-induced expression of cytokinin reverses pistil abortion during maize flower development. Plant J. 38, 910–922

Yu, J. & Ao, G. (1997) Expression of 10 kDa sulfur-rich prolamin gene of rice in potato. Acta Bot. Sin. 39, 329–334

Yu, O. et al. (2000) Production of the isoflavones genistein and daidzein in nonlegume dicot and monocot tissues. Plant Physiol. 124, 781–794

Yu, O. et al. (2003) Metabolic engineering to increase isoflavone biosynthesis in soybean seed. Phytochemistry 63, 753–763

Zeh, M. et al. (2001) Antisense inhibition of threonine synthase leads to high methionine content in transgenic potato plants. Plant Physiol. 127, 792–802

Zhang, P., Wang, W.Q., Zhang, G.L., Kaminek, M., Dobrev, P., Xu, J., & Gruissem, W. (2010) Senescence-inducible expression of isopentenyl transferase extends leaf life, increases drought stress resistance and alters cytokinin metabolism in cassava. J. Integr. Plant Biol.52, 653–669

Zhao, Z.Y.G.K. et al. (2002) Nutritionally improved transgenic sorghum. Plant Biotechnology 2002 and Beyond. Proceedings of the 10th IAPTC&B Congress, Orlando, FL, June 23–28, 2002

BOARD OF SCIENTIFIC AND POLICY ADVISORS

BOARD OF SCIENTIFIC AND POLICY ADVISORS (CONTINUED)

BOARD OF SCIENTIFIC AND POLICY ADVISORS (CONTINUED)

BOARD OF SCIENTIFIC AND POLICY ADVISORS (CONTINUED)

BOARD OF SCIENTIFIC AND POLICY ADVISORS (CONTINUED)

BOARD OF SCIENTIFIC AND POLICY ADVISORS (CONTINUED)

Made in United States
Cleveland, OH
05 November 2026

47649168R00067